MW00389563

RADIO FREQUENCY
PRINCIPLES AND APPLICATIONS

Books of Related Interest from IEEE Press...

WIRELESS COMMUNICATIONS: Principles and Practices
Theodore Rappaport
Published in cooperation with Prentice Hall
1996   Hardcover   672 pp   IEEE Order No. PC5641   ISBN 0-7803-1167-1

THE MOBILE COMMUNICATIONS HANDBOOK
Edited by Jerry D. Gibson
Published in cooperation with CRC Press
1996   Hardcover   624 pp   IEEE Order No. PC5653   ISBN 0-8493-8573-3

RADIO FREQUENCY PRINCIPLES AND APPLICATIONS

The Generation, Propagation, and Reception of Signals and Noise

Albert A. Smith, Jr.
Fellow, IEEE

IEEE
PRESS

IEEE Electromagnetic Compatibility Society, *Sponsor*

IEEE Microwave Theory and Techniques Society, *Sponsor*

IEEE Press/Chapman & Hall Publishers Series on Microwave Technology and RF

The Institute of Electrical and Electronics Engineers, Inc., New York

Printed in the United States of America

10   9   8   7   6   5   4   3   2   1

ISBN 0-7803-3431-0
IEEE Order Number: PC5704

Library of Congress Cataloging-in-Publication Data

Smith, Albert A., 1935–
    Radio frequency principles and applications : the generation,
propagation, and reception of signals and noise / Albert Smith, Jr.
        p.   cm. -- (IEEE Press / Chapman & Hall Publishers series on
microwave technology and techniques)
    Includes bibliographical references and index.
    ISBN 0-7803-3431-0 (alk. paper)
    1. Radio waves.   2. Radio--Transmitters and transmission.
3. Radio--Receivers and reception.   4. Radio--Interference.
I. Title. II. Series.
TK6550.S49795   1998
621.384 ' 11--dc21   98-6458
                                                                  CIP

To
Denise, Matthew,
and
Kaitlin

CONTENTS

Preface   xiii

Chapter 1   Static Fields and Sources   1
             1.1   Point Charge—Coulomb's Law   1
             1.2   Electric Flux Density and Gauss's Law   2
             1.3   Electric Flux   3
             1.4   Conservation of Energy   4
             1.5   Potential Difference   4
             1.6   Field from Line and Surface Charges   5
             1.7   Static E-Field Summary   6
             1.8   Line Current—Biot–Savart Law   7
             1.9   Magnetic Field from a Line Current   8
             1.10   Magnetic Flux Density and Magnetic Flux   8
             1.11   Ampere's Law   9
             1.12   Lorentz Force   10
             1.13   Magnetic Field Units and Conversions   11
             1.14   Static Magnetic Field Summary   12
                    References   12

Chapter 2   Time-varying Fields   13
             2.1   Faraday's Law   13
             2.2   Maxwell's Equations—Region
                    with Sources   15

2.3   Maxwell's Equations—Source-Free
      Region   17
2.4   Maxwell's Equations—Sinusoidal
      Fields   18
2.5   Boundary Conditions   19
2.6   Plane Wave Incident on a Conducting
      Half Space   21
2.7   Diffusion and Skin Depth   24
2.8   Transmission through a Metal Sheet
      (Shielding Effectiveness)   26
2.9   Fields from a Short Dipole   28
2.10 Fields from a Small Loop   31
2.11 Near-Field and Far-Field Regions   33
2.12 Wave Impedance   36
2.13 Power Density and Hazardous
      Radiation   38
      References   40

CHAPTER 3   PROPAGATION   43
3.1   Free-Space Propagation   43
3.2   Ground-Wave Propagation
      over Plane Earth   46
        Ground-Wave Model   49
        Attenuation and Height Gain Curves   52
3.3   Propagation over a Perfectly
      Conducting Plane   57
        Horizontal (Perpendicular) Polarization   57
        Vertical (Parallel) Polarization   58
3.4   Attenuation of Electromagnetic Fields
      by Buildings   60
3.5   Edge Diffraction   63
        Knife Edge Example   66
3.6   Rayleigh Roughness Criterion   66
      References   67

CHAPTER 4   ANTENNAS   69
4.1   Antenna Parameters   69
        Effective Length   71
        Antenna Factor   71
        Power Density   72
        Radiation Intensity   72
        Directive Gain   73
        Directivity   74
        Power Gain   74
        Realized Gain   74

Total Antenna Efficiency 75
Radiation Efficiency 75
Radiation Resistance 75
Effective Aperture 76
Maximum Effective Aperture 76
Antenna Noise Temperature 76
External Noise Factor 76
Antenna Reflection Coefficient 77
Antenna VSWR 77
4.2 Relationships Between Antenna
Parameters 77
Power Gain and Directive Gain 78
Realized Gain and Directive Gain 78
Power Gain and Realized Gain 78
Maximum Effective Aperture,
Directivity, and Gain 79
Effective Aperture, Directivity, and Gain 79
Power Density, Directivity, and Gain 80
Electric Field and Radiated Power 80
Effective Length and Directivity 81
Antenna Factor and Realized Gain 82
Antenna Factor and Power Gain 83
4.3 Reciprocity 84
4.4 Types of Receiving Antennas 87
Small Loop Antennas 87
Tunable Dipoles 89
Broadband Dipoles 90
Log-Periodic Antennas 90
Vertical Monopoles 92
Pyramidal Horns 92
Reflector Antennas 93
Discone Antennas 93
Radiation Monitor Probes 94
4.5 Antenna Calibration 94
Standard-Antenna Method 95
Standard-Field Method 97
Secondary Standard-Antenna Method 99
Standard-Site Method 101
References 108

Chapter 5 The RF Environment 111
5.1 Noise Parameters 111
Noise Power Parameters 112
Noise Field-Strength Parameters 113
Received Voltage Parameters 115
5.2 The Receiving System 116
System Noise Factor 116

		Received Noise	118
		Receiver Sensitivity and Noise Figure	120
	5.3	Extraterrestrial Noise	125
	5.4	Atmospheric Noise	127
	5.5	Man-Made Radio Noise	130
	5.6	Power-Line Conducted Noise	132
	5.7	Earth's Magnetic and Electric Fields	135
		References	138

CHAPTER 6   WAVEFORMS AND SPECTRAL ANALYSIS   141

	6.1	Classification of Signals	141
	6.2	Fourier Transform	142
	6.3	Spectral Intensity	143
		Network Response	144
		RMS Spectral Intensity	145
		Impulse Bandwidth Definition	146
		Examples	146
	6.4	Fourier Series	151
		Narrowband and Broadband Response	153
		References	155

CHAPTER 7   TRANSMISSION LINES   157

	7.1	Examples of Transmission Lines	157
	7.2	Transverse Electromagnetic (TEM) Mode of Propagation	158
	7.3	Two-Conductor Transmission Line Model	160
	7.4	Distributed Parameters	161
	7.5	Propagation Constant and Characteristic Impedance	162
		Lossless Lines	164
		Lossy Lines at Radio Frequencies	166
		Calculation of Transmission Line Constants	166
	7.6	Reflection and Transmission Coefficients	168
	7.7	Sinusoidal Steady-State Solutions	170
		Voltage and Current Distributions	170
		Voltages and Currents at the Terminations	171
		Voltage Standing-Wave Ratio (VSWR)	174
	7.8	Excitation by External Electromagnetic Fields	176
		Two-Conductor Line	177
		Conductor Over a Ground Plane	182

7.9   Radiation from Transmission Lines   188
       Two-Conductor Line   189
       Conductor Over a Ground Plane   193
       References   194

Appendix A   Physical Constants   197

Appendix B   Electrical Units   199

Appendix C   Wave Relations   201

Appendix D   Math Identities   203

Appendix E   Vector Operators   205

Appendix F   Frequency Bands   207

Index   209

About the Author   219

PREFACE

This book is intended primarily for practicing Radio Frequency (RF) engineers engaged in the prediction, analysis, and measurement of electromagnetic fields and their effects. Engineers working in the many diverse fields of RF technology, including EMC, radio wave propagation, antennas, the radio frequency environment, wireless communications and microwaves, will find much of the material to be of value. It can also serve as a useful reference for senior and graduate-level students in electromagnetics, RF design, EMC, communication systems, wave propagation, antennas, and transmission lines.

The emphasis in this book is on the practical solution of problems involving the generation, propagation, and reception of electromagnetic signals and noise. The material is derived from a variety of sources, including journal articles, symposium papers, textbooks, government research reports, publications of international technical committees, standards publications, and my own research and notes. My goal was to extract the essential information from these sources and to present the material in a readily understandable and usable form. The focus is on applications. Abstruse theory, complex mathematical derivations and intricate experimental details were purposely avoided. Ample references are provided for those who wish to pursue the subject matter in more detail.

The material is organized as follows. Chapter 1 is a review of the basic laws which govern static electric and magnetic fields. The relations between static fields and their sources are examined, including Coulomb's law, Gauss's law, the Biot-Savart law, Ampere's law, and Lorentz forces. Chapter 2 begins with a review of the fundamental laws of time-varying fields—Faraday's law and Maxwell's equations. Most of this chapter is devoted to practical applications including boundary conditions, reflection and transmission of plane waves at a metal surface, diffusion and skin depth, shielding effectiveness, the fields from a short dipole and a small loop, near-field and far-field regions, wave impedance, and power density and hazardous radiation. The subject of Chapter 3 is radio wave propagation. Topics covered are free-space propagation, ground-wave propagation over plane earth, propagation over a perfectly conducting plane, building attenuation, edge diffraction, and the Rayleigh roughness criterion. Chapter 4 is a review of the important characteristics of transmitting and receiving antennas. Common antenna parameters are defined. Reciprocity relationships between transmitting and receiving antennas are reviewed. The types of antennas commonly used to measure electromagnetic fields are described. Last, methods for calibrating receiving antennas are described in detail. The subject of Chapter 5 is the radio frequency environment. Common noise parameters are reviewed. The response of a receiving system to external noise sources is studied. Receiver sensitivity and noise figure are reviewed. Extraterrestrial noise, atmospheric noise, man-made radio noise and power-line conducted noise spectra are presented. Data on the earth's static magnetic and electric fields are provided. Chapter 6, on waveforms and spectral analysis, reviews the Fourier transform and spectral intensity of deterministic energy signals, and the Fourier series of periodic power signals. Chapter 7 is a review of two-conductor transmission line theory, including the excitation of transmission lines by external electromagnetic fields, and radiation from differential-mode and common-mode currents on transmission lines. Appendixes provide useful information on physical constants, electrical units, wave relations, math identities, vector operators, and frequency bands.

Many colleagues have contributed to the author's knowledge of electromagnetics over the years, especially Edwin L. Bronaugh, Donald N. Heirman, Motohisa Kanda, Richard B. Schulz, Ralph M. Showers, Edward N. Skomal, and Harold E. Taggart. I would also like to thank my associates at the IBM Corporation who provided support and encouragement on many complex projects, including Donald R. Bush, Ralph Calcavecchio, Robert F. German, George Holzman, Ulrich Kirste, William F. McCarthy, Carl H. Meyer, James B. Pate, Larry J. Prescott, Edwin R. Rose, and Chang-Yu Wu. The author owes a special debt of gratitude to Professor Clayton R. Paul, University of Kentucky at Lexington, for his many helpful insights on the coupling of fields to transmission lines.

Albert A. Smith, Jr.
Woodstock, New York
1997

CHAPTER 1

STATIC FIELDS AND SOURCES

This chapter is a brief review of the basic laws governing static electric and magnetic fields. The defining relations between static fields and their sources are examined. Definitions and units of measurements for electric and magnetic field strength, flux, flux density, and force are summarized. For a more comprehensive study of static electric and magnetic fields, refer to any elementary text on electromagnetics such as Ramo, Whinnery, and Van Duzer [1]; Kraus and Carver [2]; Plonsey and Collin [3]; or Paul and Nasar [4].

1.1 POINT CHARGE—COULOMB'S LAW

Charles Augustin de Coulomb (1736–1806) was a French physicist, inventor and army engineer. He made many fundamental contributions in the fields of friction, electricity and magnetism, including the formulation of *Coulomb's law*. The unit for electric charge was named in his honor.[1]

The source of the static electric field is stationary charge. The simplest source is a point charge Q as shown in Fig. 1.1. If a unit positive test charge q is placed in the vicinity of Q, a force $\mathbf{F}$ is exerted on the test charge which is given by Coulomb's law as

$$\mathbf{F} = \frac{Qq}{4\pi \varepsilon R^2} \mathbf{a}_R \tag{1.1}$$

[1] Much of the biographical data in Chapters 1 and 2 is from the *World Book Encyclopedia*, Free Enterprises Educational Corp., Chicago, 1973.

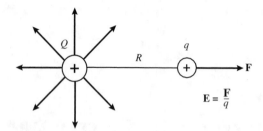

Figure 1.1 Radial electric field from a point charge.

where **F**   force in newtons, N
     Q charge in coulombs, C
     q unit positive test charge, one coulomb
     ε $= \varepsilon_r \varepsilon_o$ permittivity of the medium, farads/meter (F/m)
     ε_r relative permittivity (or dielectric constant) of the medium
     ε_o permittivity of free space, 8.854×10^{-12} farads/meter
     R distance between charges in meters (m)
     $\mathbf{a}_R$ unit vector in the radial direction.

By definition, the electric field strength **E** is

$$\mathbf{E} \equiv \frac{\mathbf{F}}{q} \qquad \text{newtons per coulomb (N/C).} \qquad (1.2)$$

For a point charge then,

$$\mathbf{E} = \frac{Q}{4\pi \varepsilon R^2}\mathbf{a}_R \qquad \text{volts per meter (V/m).} \qquad (1.3)$$

In the International System of Units (SI), the units of volts per meter and newtons per coulomb are equivalent.

Note that the radial electric field from a point charge falls off as $1/R^2$.

1.2   ELECTRIC FLUX DENSITY AND GAUSS'S LAW

Karl Friedrich Gauss (1777–1855) was a German mathematician. Often referred to as the Prince of Mathematics, he is considered one of the greatest mathematicians of all time, ranked with Archimedes and Newton. A child prodigy, he became famous for his work in number theory, geometry, astronomy, and for important contributions to the mathematical theory of electromagnetism. His inventions include the bifilar magnetometer and the electric telegraph.

Electric field strength **E** has dimensions of volts per meter and is a measure of the *intensity* of the field.

Electric flux density **D** is defined as

$$\mathbf{D} = \varepsilon\mathbf{E} \tag{1.4}$$

and has dimensions of coulombs/m^2, or charge per unit area. Thus the term flux *density*. **D** is sometimes called the electric displacement vector.

While **E** is dependent on the permittivity of the medium, **D** is independent of the medium (assuming the medium is isotropic) and depends only on the sources of charge.

For the point charge in Fig. 1.1, we have from (1.3) and (1.4)

$$\mathbf{D} = \frac{Q}{4\pi R^2}\mathbf{a}_R \qquad \text{coulombs per square meter (C/m}^2\text{).}$$

Now consider the more general case of a distribution of charges.

This could be a distribution of point charges q_i, a line charge distribution with density ρ_l, a surface charge distribution with density ρ_s, a volume charge distribution with density ρ_v, or a combination of these. If these sources are contained within an arbitrary closed surface S as indicated in Fig. 1.2, Gauss's law states that

$$\oint_S \mathbf{D} \cdot d\mathbf{S} = Q_{\text{enclosed}}. \tag{1.5}$$

That is, the integral of the normal component of the electric flux density over any closed surface is equal to the net charge enclosed.

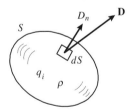

Figure 1.2 Illustration of Gauss's law.

The charge enclosed is given by, in general,

$$Q_{\text{enclosed}} = \sum q_i + \int_l \rho_l dl + \int_s \rho_s\, ds + \int_v \rho_v\, dv.$$

1.3   ELECTRIC FLUX

The electric flux ψ passing through a surface **S** is defined as the product of the normal flux density D_n and the surface area S, assuming that **D** is uniform over S. More generally, when $\mathbf{D}$ is not uniform over the surface (Fig. 1.3), ψ is the surface integral of the scalar product of $\mathbf{D}$ and $d\mathbf{S}$, or

$$\psi = \int_S \mathbf{D} \cdot d\mathbf{S} \quad \text{coulombs (C)}.$$

An interesting aside—The surface exists in space and contains no charges. If ψ is a time-varying function, the quantity $d\psi/dt$ has dimensions of coulombs per second, or amperes. That is, a time rate of change of electric flux is associated with a current. This is called displacement current, a concept introduced by Maxwell.

Figure 1.3 Electric flux density.

1.4 CONSERVATION OF ENERGY

If a test charge is moved around any closed path in a static electric field, no net work is done. Since the charge returns to its starting point, the forces encountered on one part of the path are exactly offset by opposite forces on the remainder of the path. The mathematical statement for conservation of energy in a static electric field is

$$\oint \mathbf{E} \cdot d\mathbf{l} = 0. \tag{1.6}$$

This statement is not true for time varying fields, in which case Faraday's law applies.

1.5 POTENTIAL DIFFERENCE

The potential difference between two points a and b immersed in an electric field E (see Fig. 1.4) is defined as the work required to move a unit positive test charge from a to b and is given by

$$V_{ab} = - \int_a^b \mathbf{E} \cdot d\mathbf{l} \quad \text{joules per coulomb (J/C) or volts (V).} \tag{1.7}$$

The potential difference is independent of the path taken from a to b. That is, it depends only on the endpoints. The negative sign in (1.7)

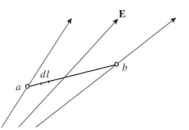

Figure 1.4 Potential difference.

indicates that the field does work on the positive charge in moving from a to b and there is a fall in potential (negative potential difference).

1.6   FIELD FROM LINE AND SURFACE CHARGES

Refer to Fig. 1.5a. The radial electric field from a uniform line charge of infinite extent is

$$E_R = \frac{\rho_l}{2\pi \varepsilon_o R} \qquad (1.8)$$

where E_R radial electric field strength, V/m
      ρ_l line charge density, coulombs/m
      R radial distance, m
      ε_o permittivity of free space.

The transverse component of the electric field is zero. Note that the radial component falls off as $1/R$. This solution has applications in transmission line problems, including overhead power distribution and transmission lines.

The field normal to a surface of infinite extent having a uniform surface charge density ρ_s (Fig. 1.5b) is

$$E_n = \frac{\rho_s}{\varepsilon_o} \qquad (1.9)$$

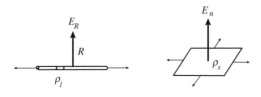

(a) Line charge   (b) Surface charge

Figure 1.5 Line and surface charges.

and

$$D_n = \rho_s \qquad (1.10)$$

where E_n normal electric field, V/m
$\quad\quad D_n$ normal flux density, coulombs/m^2
$\quad\quad \rho_s$ surface charge density, coulombs/m^2.

The field above the surface is constant because the surface is infinite in extent. Again, there is no transverse field component.

1.7   STATIC E-FIELD SUMMARY

A summary of important static electric field relations is provided in Table 1.1. The electric fields from a point charge, an infinite line charge, and an infinite surface charge are given in Table 1.2.

Table 1.1 Summary of Static Electric Field Relations

	Definition	Units
Coulomb's law	$\mathbf{F} = \dfrac{Qq}{4\pi\varepsilon R^2}\mathbf{a}_R$	newtons (N)
E Field	$\mathbf{E} \equiv \dfrac{\mathbf{F}}{q}$	V/m or N/C
Flux density	$\mathbf{D} = \varepsilon\mathbf{E}$	C/m^2
Gauss's law	$\oint_S \mathbf{D}\cdot d\mathbf{S} = Q_{\text{enclosed}}$	C
Electric flux	$\psi = \int_S \mathbf{D}\cdot d\mathbf{S}$	C
Potential difference	$V_{ab} = -\int_a^b \mathbf{E}\cdot d\mathbf{l}$	V or J/C

Table 1.2 Fields from Various Sources

Source	E Field
Point charge	$\mathbf{E} = \dfrac{Q}{4\pi\varepsilon R^2}\mathbf{a}_R$
Infinite line charge	$E_R = \dfrac{\rho_l}{2\pi\varepsilon_o R}$
Infinite surface charge	$E_n = \dfrac{\rho_s}{\varepsilon_o}$

1.8   LINE CURRENT—BIOT-SAVART LAW

> Biot and Savart, in 1820, established the basic experimental laws relating
> magnetic field strength to electric currents. They also established the law of
> force between two currents.

The source of static magnetic fields is charge moving at a constant
velocity, namely, direct current (also referred to as steady current or stationary current). By definition, the current through any cross-sectional area
is equal to the time rate at which electric charge passes through the area,
or

$$I = \frac{dq}{dt} \qquad \text{coulombs per second (C/sec) or amperes (A).} \qquad (1.11)$$

The current gives rise to a magnetic field. For example, consider an infinitesimal current element $I\,d\mathbf{l}$ as shown in Fig. 1.6. I is the magnitude
of the current element, and $d\mathbf{l}$ is a unit vector that defines the direction.

The differential magnetic field strength $d\mathbf{H}$ in vector notation is given
by

$$d\mathbf{H} = \frac{I\,d\mathbf{l} \times \mathbf{a}_R}{4\pi R^2} \qquad \text{amperes per meter (A/m)} \qquad (1.12)$$

and is one form of the Biot-Savart law. (This law is sometimes attributed
to Ampere.)

The magnitude of (1.12) is

$$dH = \frac{I\,dl\,\sin\theta}{4\pi R^2} \qquad \text{A/m.} \qquad (1.13)$$

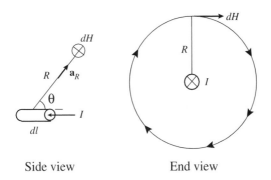

Side view   End view

Figure 1.6 Illustration of the Biot-Savart law.

1.9   MAGNETIC FIELD FROM A LINE CURRENT

The magnetic field strength H_ϕ from a line current of infinite extent (Figure 1.7) is obtained directly from the Biot-Savart law by integrating (1.12) over the length of the conductor. The result is

$$H_\phi = \frac{I}{2\pi R} \qquad \text{A/m.}$$

The magnetic field strength from an infinite line current falls off as $1/R$. This solution has applications in transmission line analysis, including overhead powerlines.

The direction of the H-field vector in relation to the direction of current flow is given by the "right-hand rule." See Figs. 1.6 and 1.7. If the thumb points in the direction of the current flow, the fingers indicate the positive direction of the magnetic field strength.

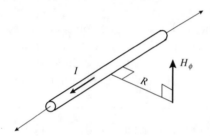

Figure 1.7 Magnetic field from an infinite line current.

1.10   MAGNETIC FLUX DENSITY AND MAGNETIC FLUX

Magnetic field strength **H** has dimensions of amperes per meter and is a measure of the *intensity* of the field, analogous to the electric field strength **E**.

Magnetic flux density **B** is defined as

$$\mathbf{B} = \mu\mathbf{H} \qquad (1.14)$$

where **B**   magnetic flux density, webers/m^2 (Wb/m^2) or tesla (T)
       μ $= \mu_r\mu_o$ permeability of the medium, henrys/m (H/m)
       μ_r relative permeability of the medium
       μ_o permeability of free space, $4\pi \times 10^{-7}$ henrys/meter
       H   magnetic field strength, A/m.

B has dimensions of weber/m^2, or flux per unit area, and is therefore called flux *density*. **B** is analogous to electric flux density **D**.

Lines of magnetic flux are conceptually similar to lines of electric flux (except that lines of magnetic flux close on themselves while electric flux lines terminate on charges). The magnetic flux Φ passing through a surface **S** is defined as the product of the normal magnetic flux density B_n and the surface area S. This assumes that **B** is uniform over **S**. More generally, when **B** is not uniform over the surface (see Fig. 1.8), the magnetic flux is given by the integral over the surface of the scalar product of **B** and $d\mathbf{S}$, that is,

$$\Phi = \int_S \mathbf{B} \cdot d\mathbf{S} \qquad \text{webers (Wb).} \qquad (1.15)$$

(Note that the symbol used for magnetic flux is the uppercase Greek Φ to distinguish it from the spherical coordinate denoted by the lowercase Greek ϕ.) The unit of webers is equivalent to volt-seconds.

Figure 1.8 Magnetic flux density.

1.11   AMPERE'S LAW

> Andre Marie Ampere (1775–1836) was a French mathematician and physicist. His experiments led to the law of force between current carrying conductors and to the invention of the galvanometer. He postulated that magnetism was due to circulating currents on an atomic scale, showing the equivalence of magnetic fields produced by currents and those produced by magnets.

Ampere's law states that the line integral of the tangential magnetic field strength around any closed path is equal to the net current enclosed by that path. Mathematically,

$$\oint \mathbf{H} \cdot d\mathbf{l} = I_{\text{enclosed}}. \qquad (1.16)$$

For example, the closed integral of the tangential H-field around the circular path in Fig. 1.6 yields the current I. However, Ampere's law is more general in that it applies to any closed path and it applies to magnetic fields arising from both conduction and convection currents. (Convection currents are charges or charge densities moving with velocity v, i.e., qv and ρv.) For time-varying fields, the right-hand side of (1.16) contains an additional displacement current term.

Ampere's law is the magnetic field analog of Gauss's law, except that it involves a closed contour rather than a closed surface. Comparing (1.16) with (1.6) reveals that while the static electric field is a conservative field, the magnetic field is not.

1.12   LORENTZ FORCE

Hendrick Antoon Lorentz (1853–1928) was a Dutch physicist who became famous for his electron theory of matter. In addition to the formulation of the *Lorentz force* equation, he developed the *Lorentz transformations*, which show how bodies are deformed by motion, and the *Lorentz condition*, which has special significance in relativistic field theory. He shared the 1902 Nobel prize for physics with Pieter Zeeman for discovering the Zeeman effect of magnetism on light.

A point charge q moving with a velocity $\mathbf{v}$ in a magnetic field $\mathbf{B}$ experiences a force, called the Lorentz force, which is given by

$$\mathbf{F} = q\mathbf{v} \times \mathbf{B} \qquad \text{newtons (N).} \qquad (1.17)$$

Equation (1.17) assumes that there is no electric field acting on the point charge. If an electric field is present, the total force acting on the point charge is the sum of the Lorentz force and the $q\mathbf{E}$ force from (1.2), that is,

$$\mathbf{F} = q(\mathbf{E} + \mathbf{v} \times \mathbf{B}) \qquad \text{newtons (N).} \qquad (1.18)$$

The Lorentz force on an infinitesimal current element $I\,d\mathbf{l}$ immersed in a magnetic field $\mathbf{B}$ follows from (1.17), the definition of current $I = dq/dt$, and the velocity $\mathbf{v} = d\mathbf{l}/dt$. We have

$$dq\,\mathbf{v} = dq\,\frac{d\mathbf{l}}{dt} = \frac{dq}{dt}\,d\mathbf{l} = I\,d\mathbf{l}.$$

The instantaneous force on the infinitisimal current element is

$$d\mathbf{F} = I\,d\mathbf{l} \times \mathbf{B} \qquad \text{newtons (N).} \qquad (1.19)$$

For the special case of a linear conductor of length $\mathbf{L}$ carrying a current I in a static uniform magnetic field $\mathbf{B}$ (Fig. 1.9), the Lorentz force is

$$\mathbf{F} = I\mathbf{L} \times \mathbf{B} \quad \text{newtons (N)} \qquad (1.20)$$

and the magnitude is

$$F = ILB\sin\theta. \qquad (1.21)$$

The Lorentz force is proportional to the magnitude of the current, the length of the conductor, the strength of the magnetic field and the sine of the angle

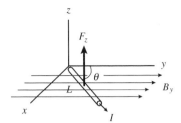

Figure 1.9 Lorentz force on a conductor in a uniform magnetic field.

between the current and the field. The direction of the force is perpendicular to the plane containing the current and the magnetic field.

The Lorentz force is utilized in Hall effect devices, in the focusing and deflection of electron beams in cathode-ray tubes, and in galvanometer movements, to mention a few applications. The Lorentz force has also been suggested as a possible factor in the biological effects of electromagnetic fields. In particular, the movement of charged electrolytes in body fluids in a magnetic field (for example, the earth's magnetic field) may cause changes in body chemistry.

Equation (1.20) can also be interpreted as the defining relation for the magnetic field **B**. If $I\,d\mathbf{l}$ and **B** are perpendicular

$$B = \frac{F}{I\,dl}. \tag{1.22}$$

That is, the magnetic flux density can be defined as a force per unit current element, analogous to the definition of the electric field as a force per unit charge in equation (1.2). The units of B are the Tesla, or equivalently, webers/m^2 or newtons/ampere-meter.

1.13 MAGNETIC FIELD UNITS AND CONVERSIONS

While SI units are preferred for magnetic field quantities, many applications still use centimeter-gram-second (cgs) units as a matter of custom or tradition. Table 1.3 is a summary of SI and cgs magnetic field units and the conversion factors from one system of units to the other.

ELF magnetic fields from appliances, video display terminals, and power distribution and transmission lines are commonly measured in milligauss. DC magnetic field measurements of magnets and magnetized objects (for complying with FAA regulations for air shipments, for example) are also commonly expressed in milligauss. In fact, most of the instruments used for DC and ELF magnetic field measurements are calibrated in milligauss.

Table 1.3 MAGNETIC FIELD UNITS

	Quantity			
	H	**B**	**Φ**	**μ_o**
SI units	Amps/meter A/m	Tesla T	Weber Wb	$4\pi \times 10^{-7}$ H/m
cgs units	Oersted Oe	Gauss Gs	Maxwell Mx	1
Conversion	1 Oe = 79.6 A/m	1 Gs = 10^{-4}T	1 Mx = 10^{-8}Wb	

1.14 STATIC MAGNETIC FIELD SUMMARY

A summary of important static magnetic field relations is provided in Table 1.4.

Table 1.4 SUMMARY OF STATIC MAGNETIC FIELD RELATIONS

	Definition	Units
Biot–Savart law	$d\mathbf{H} = \dfrac{I d\mathbf{l} \times \mathbf{a}_R}{4\pi R^2}$	A/m
Flux density	$\mathbf{B} = \mu \mathbf{H}$	T
Magnetic flux	$\Phi = \int_S \mathbf{B} \cdot d\mathbf{S}$	Wb
Ampere's law	$\oint \mathbf{H} \cdot d\mathbf{l} = I_{\text{enclosed}}$	A
Lorentz force	$\mathbf{F} = I\mathbf{L} \times \mathbf{B}$	N

REFERENCES

[1] S. Ramo, J. R. Whinnery, and T. Van Duzer, *Fields and Waves in Communications Electronics*, third edition, John Wiley and Sons, New York, Chichester, Brisbane, Toronto, 1994.

[2] J. D. Kraus and K. R. Carver, *Electromagnetics*, second edition, McGraw-Hill, New York, 1973.

[3] R. Plonsey and R. E. Collin, *Principles and Applications of Electromagnetic Fields*, second edition, McGraw-Hill, New York, 1982.

[4] C. R. Paul and S. A. Nasar, *Introduction to Electromagnetic Fields*, second edition, McGraw-Hill, New York, 1987.

CHAPTER 2

TIME-VARYING FIELDS

This chapter begins with a review of the fundamental laws which govern time-varying electromagnetic fields, including Faraday's law, Maxwell's equations, and boundary conditions at the interface between regions with different electrical properties. The remainder of the chapter is devoted to more practical applications of the theory, including reflection and transmission at a metal surface, diffusion and skin depth, the fields from a short dipole and a small loop, near-field and far-field regions, wave impedance, and power density.

2.1 FARADAY'S LAW

Michael Faraday (1791–1867) was one of the greatest English chemists and physicists. He discovered the principle of electromagnetic induction on August 29, 1831. The electric generator and electric motor are based on Faraday's law. The first evidence for the existence of the electron dates to the work of Faraday. He worked and lectured at the Royal Institution in London for 54 years.

In 1831, Michael Faraday found experimentally that a time-changing magnetic field induced a voltage in a coil immersed in the field, and that the voltage was equal to the time rate of change of the magnetic flux linking the coil. See Fig. 2.1. This form of Faraday's law of induction is expressed as

$$V = -\frac{d\Phi}{dt} \qquad (2.1)$$

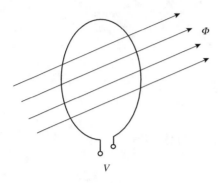

Figure 2.1 Voltage induced in a wire loop by a changing magnetic field (Faraday's law).

where V induced voltage, volts
       Φ magnetic flux, webers.

The minus sign in (2.1) follows from Lenz's law which states that the direction of the induced voltage is such that it will tend to produce a current flow that opposes the change of flux.

An important discovery that evolved from Faraday's experiments was that a time-changing magnetic field produces an electric field. This holds in any medium—free space, metals, dielectrics, etc. This is the basis for electromagnetic wave propagation in free space where electric and magnetic fields produce each other in the absence of any sources or a material medium. It is also the basis for one of Maxwell's equations.

Consider any closed path C in space as in Fig. 2.2. Defining the induced voltage V in (2.1) as the line integral of the electric field around C, and using (1.15) to define Φ , results in the following integral form of Faraday's law:

$$\oint_C \mathbf{E} \cdot d\mathbf{l} = -\frac{\partial}{\partial t} \int_S \mathbf{B} \cdot d\mathbf{S}. \qquad (2.2)$$

This is also known as the integral form of Maxwell's equation. In a static field, the right-hand side of (2.2) is zero. See (1.6).

A practical application of Faraday's law is the calculation of the voltage induced in an N -turn loop antenna excited by an incident H -field. Refer to Fig. 2.3. In this example, the magnetic field strength H is uniform and normal to the plane of the loop, and has a sinusoidal time variation $\varepsilon^{j\omega t}$. The area of the loop antenna is S . The shape of the loop is of no consequence. It could be square as shown in Fig. 2.3, or circular, triangular, polygonal, etc.

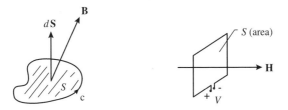

Figure 2.2 Integral form of Faraday's law.   **Figure 2.3** Loop antenna.

The received voltage from (2.1) and (2.2) is

$$V = -N\frac{d\Phi}{dt} = -N\frac{d}{dt}\int_S \mathbf{B} \cdot d\mathbf{S}$$

$$V = -\mu_o N S \frac{d}{dt} H \varepsilon^{j\omega t}$$

or

$$V = -j\omega\mu_o N S H \varepsilon^{j\omega t} \qquad \text{V}. \tag{2.3}$$

Equation (2.3) is valid for frequencies where the perimeter or circumference of the loop is less than approximately 0.1 wavelength.

2.2   MAXWELL'S EQUATIONS—REGION WITH SOURCES

James Clerk Maxwell (1831–1879), a Scottish scientist, was one of the greatest mathematicians and physicists of the nineteenth century. His best known work, *A Treatise on Electricity and Magnetism* (Oxford University Press, 1873; 3rd edition, 1904) is the foundation of present-day electromagnetic theory. He formulated exact mathematical descriptions of electric and magnetic fields based on Faraday's experiments. In 1864, Maxwell predicted the existence of electromagnetic waves that travel through space at the speed of light. The German physicist Heinrich Hertz proved Maxwell's theories between 1886 and 1888 when he produced electromagnetic waves using an oscillating current generated by the spark of an induction coil.

Consider a region with a charge density ρ and a current density $\mathbf{J}$. This region is characterized by its physical constants of conductivity σ, permeability μ, and permittivity ε, as illustrated in Fig. 2.4.

The presence of an electric field $\mathbf{E}$ in a region with conductivity σ will produce a *conduction current density*

$$\mathbf{J}_{\text{COND}} = \sigma \mathbf{E} \qquad \text{amps/m}^2 \tag{2.4}$$

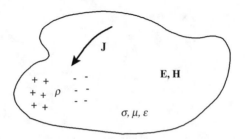

Figure 2.4 Region with sources.

where σ conductivity, mhos/meter
     E   electric field strength, V/m.

Equation (2.4) is the field-theoretic form of ohm's law.

In a nonconducting region, a charge density ρ moving with velocity **v** is the equivalent of current density. Examples are electron beams in cathode ray tubes and electron streams in ionized gases and plasmas. This equivalent current density is called *convection current density* and is defined as

$$\mathbf{J}_{\text{CONV}} = \rho\mathbf{v} \qquad \text{amps/m}^2 \tag{2.5}$$

where ρ charge density, C/m^3
     v   velocity, m/sec.

Total current density **J** is the sum of conduction and convection current densities:

$$\mathbf{J} = \mathbf{J}_{\text{COND}} + \mathbf{J}_{\text{CONV}}. \tag{2.6}$$

Current and charge are related by the continuity equation

$$\nabla \cdot \mathbf{J} = -\frac{\partial\rho}{\partial t} \tag{2.7}$$

which states that the net flow of current out of a volume is equal to the negative time rate of change of the volume charge density (or the rate of decrease of charge).

In the formulation of his equations for the electromagnetic field, Maxwell proposed that in a time-varying field, the time rate of change of the electric flux density is equivalent to a current density and produces a magnetic field similar to that produced by a conduction current density (moving charges in conductors) and a convection current density (moving charges in space). This was alluded to in Section 1.3, *Electric Flux*. This quantity

$$\frac{\partial \mathbf{D}}{\partial t} \quad \text{amps/m}^2 \tag{2.8}$$

is called *displacement current density*.

The sum of the conduction, convection, and displacement current densities in (2.6) and (2.8) appears in the right-hand side of Maxwell's equation for the curl of the magnetic field.

Maxwell's equations in derivative form for a region with sources $\mathbf{J}$ and ρ (illustrated in Fig. 2.4) are

$$\nabla \times \mathbf{E} = -\frac{\partial \mathbf{B}}{\partial t} \tag{2.9}$$

$$\nabla \cdot \mathbf{D} = \rho \quad \text{or} \quad \nabla \cdot \mathbf{E} = \rho/\varepsilon \tag{2.10}$$

$$\nabla \times \mathbf{H} = \mathbf{J} + \frac{\partial \mathbf{D}}{\partial t} \tag{2.11}$$

$$\nabla \cdot \mathbf{B} = 0 \tag{2.12}$$

$$\nabla \cdot \mathbf{J} = -\frac{\partial \rho}{\partial t} \tag{2.13}$$

Equation (2.9) is the differential form of Faraday's law of induction, (2.10) is the differential form of Gauss's law, (2.11) is the differential form of Ampere's law with the displacement current density term added, (2.12) is a consequence of the closed-loop nature of magnetic flux lines (the magnetic field has no divergence), and (2.13) is the continuity equation. (Vector operators in rectangular, cylindrical, and spherical coordinates are reviewed in Appendix E.)

2.3   MAXWELL'S EQUATIONS—SOURCE-FREE REGION

In a source-free region of space, $\sigma = 0$, $\rho = 0$, and therefore $\mathbf{J} = \mathbf{J}_{\text{COND}} + \mathbf{J}_{\text{CONV}} = 0$. Maxwell's equations in a source-free region reduce to the following form:

$$\nabla \times \mathbf{E} = -\mu \frac{\partial \mathbf{H}}{\partial t} \tag{2.14}$$

$$\nabla \cdot \mathbf{D} = \nabla \cdot \mathbf{E} = 0 \tag{2.15}$$

$$\nabla \times \mathbf{H} = \varepsilon \frac{\partial \mathbf{E}}{\partial t} \tag{2.16}$$

$$\nabla \cdot \mathbf{B} = 0. \tag{2.17}$$

If the source-free region is free space, $\varepsilon = \varepsilon_0$ and $\mu = \mu_0$.

2.4   MAXWELL'S EQUATIONS—SINUSOIDAL FIELDS

In most applications, the vector electric and magnetic fields have a sinu-soidal time variation and it is advantageous to represent the fields using the complex exponential form $e^{j\omega t}$ (or phasor notation). Thus, $\mathbf{E} = \mathbf{E}_0 e^{j\omega t}$ and $\mathbf{H} = \mathbf{H}_0 e^{j\omega t}$

where $e^{j\omega t} = \quad \cos \omega t + j \sin \omega t$
$\qquad \omega = 2\pi f \quad$ radian frequency
$\qquad f \qquad\quad$ frequency in Hz.

The instantaneous value of the field is given by the imaginary part of $\mathbf{E}_0 \, e^{j\omega t}$, that is,

$$\mathbf{E}(t) = \mathbf{E}_0 \, \text{Im} \, e^{j\omega t} = \mathbf{E}_0 \, \sin \omega t.$$

When the sinusoidal fields are expressed in complex exponential form, all time derivatives can be replaced by $j\omega$. To illustrate, if $\mathbf{F} = \mathbf{F}_0 e^{j\omega t}$ is a vector field

$$\frac{\partial}{\partial t}\mathbf{F} = \frac{\partial}{\partial t}\mathbf{F}_0 e^{j\omega t} = j\omega \mathbf{F}_0 \, e^{j\omega t} = j\omega \mathbf{F}.$$

In a region with sources ρ and $\mathbf{J}$ (Fig. 2.4), Maxwell's equations for sinu-soidal time-varying fields reduce to

$$\nabla \times \mathbf{E} = -j\omega\mu\mathbf{H} \qquad (2.18)$$

$$\nabla \cdot \mathbf{E} = \rho/\varepsilon \qquad (2.19)$$

$$\nabla \times \mathbf{H} = \sigma\mathbf{E} + j\omega\varepsilon\mathbf{E} \qquad (2.20)$$

$$\nabla \cdot \mathbf{B} = 0 \qquad (2.21)$$

$$\nabla \cdot \mathbf{J} = -j\omega\rho. \qquad (2.22)$$

In (2.20) it is assumed that the convection current density is zero, i.e., only the conduction current density $\mathbf{J} = \sigma\mathbf{E}$ exists.

In a source-free region of space where $\sigma = 0$, $\rho = 0$, and $\mathbf{J} = 0$, Maxwell's equations for sinusoidal time-varying fields are

$$\nabla \times \mathbf{E} = -j\omega\mu\mathbf{H} \qquad (2.23)$$

$$\nabla \cdot \mathbf{E} = 0 \qquad (2.24)$$

$$\nabla \times \mathbf{H} = j\omega\varepsilon\mathbf{E} \qquad (2.25)$$

$$\nabla \cdot B = 0. \qquad (2.26)$$

2.5   BOUNDARY CONDITIONS

In this section, we summarize the boundary conditions on the normal and tangential field components at the interface separating regions with different electrical properties.

Refer to Fig. 2.5, which shows the boundary between two regions with different permittivities ε, permeabilities μ, and conductivities σ. The following relations hold for the tangential components of the fields in the absence of a surface current on the boundary surface:

$$E_1^t = E_2^t \quad \text{or} \quad \varepsilon_2 D_1^t = \varepsilon_1 D_2^t \tag{2.27}$$

$$H_1^t = H_2^t \quad \text{or} \quad \mu_2 B_1^t = \mu_1 B_2^t \tag{2.28}$$

where the subscripts 1 and 2 refer to regions 1 and 2, and superscript t denotes the tangential field.

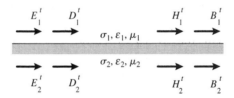

Figure 2.5 Tangential fields at boundary of two regions.

Thus the tangential components of **E** and **H** are continuous across the surface separating two media.

If a surface current J_s exists on the boundary surface, the tangential component of **H** is discontinuous by an amount equal to the surface current:

$$H_1^t - H_2^t = J_s. \tag{2.29}$$

Refer to Fig. 2.6. The following relations hold for the normal components of the fields in the absence of a charge density on the boundary surface:

$$D_1^n = D_2^n \quad \text{or} \quad \varepsilon_1 E_1^n = \varepsilon_2 E_2^n \tag{2.30}$$

$$B_1^n = B_2^n \quad \text{or} \quad \mu_1 H_1^n = \mu_2 H_2^n. \tag{2.31}$$

Thus the normal components of **D** and **B** are continuous at the boundary between two media.

If a surface charge density ρ_s exists on the boundary surface, the normal component of **D** is discontinuous by an amount equal to the surface charge density:

$$D_1^n - D_2^n = \rho_s. \tag{2.32}$$

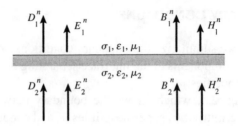

Figure 2.6 Normal field components at boundary of two regions.

The boundary conditions at the surface of a perfect conductor ($\sigma = \infty$) are of special interest. Since the conductivity of metals is so high, in many practical cases they are well approximated by a perfect conductor. An example is a field incident on a metal plane at radio frequencies. This important application will be examined in more detail in the following sections on the reflection and transmission at a conducting slab and on skin depth. Figure 2.7 shows the boundary between a perfect conductor and a dielectric medium. The unit vector normal to the surface is $\mathbf{n}$. In a perfect conductor, $E = H = 0$. Also, since the conductor is perfect ($\sigma = \infty$), the skin depth is zero and a true surface current $\mathbf{J}_s$ (in amperes per meter) exists.

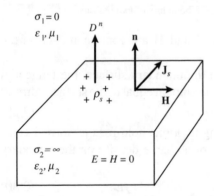

Figure 2.7 Boundary conditions at the surface of a perfect conductor.

Then, from (2.27), (2.29), (2.31) and (2.32), it follows that the fields on the surface of a perfect conductor are

$$E^t = 0 \qquad \text{or} \qquad \mathbf{n} \times \mathbf{E} = 0 \qquad\qquad (2.33)$$

$$H^t = J_s \qquad \text{or} \qquad \mathbf{n} \times \mathbf{H} = \mathbf{J}_s \qquad\qquad (2.34)$$

$$B^n = 0 \qquad \text{or} \qquad \mathbf{n} \cdot \mathbf{B} = 0 \qquad\qquad (2.35)$$

$$D^n = \rho_s \qquad \text{or} \qquad \mathbf{n} \cdot \mathbf{D} = \rho_s. \qquad\qquad (2.36)$$

Note that the surface current J_s is equal in magnitude to the tangential magnetic field H^t at the surface, and that the direction of J_s is perpendicular to H^t. This is the basis for the operation and calibration of surface current probes. The typical surface current probe is a calibrated rectangular magnetic loop antenna. When oriented for maximum response, the direction of the surface current is indicated by the transverse dimension of the probe.

2.6   PLANE WAVE INCIDENT ON A CONDUCTING HALF SPACE

Consider a plane wave normally incident on a conducting half space ($z > 0$) as illustrated in Fig. 2.8. Part of the wave is reflected from the surface of the conductor and part of the wave is transmitted into the conductor. (The transmitted wave is also called the *refracted wave*.) This is denoted as follows for the electric field component (the superscripts i, r, and t refer to the incident, reflected, and transmitted waves, respectively):

$$E_x^r = R E_x^i \qquad \text{at the surface } (z = 0) \qquad (2.37)$$

$$E_x^t = T E_x^i \qquad \text{at the surface } (z = 0) \qquad (2.38)$$

where [2] $\quad R \qquad\qquad$ reflection coefficient
$\qquad\qquad T = 1 + R \quad$ transmission coefficient.

The wave impedance of the incident plane wave in air (free space) is

$$Z_o = \frac{E^i}{H^i}. \qquad (2.39)$$

The wave impedance of the reflected plane wave in air (traveling in the negative z direction) is [1]

$$-Z_o = \frac{E^r}{H^r}. \qquad (2.40)$$

The wave impedance of the plane wave in metal is

$$Z_m = \frac{E^t}{H^t}. \qquad (2.41)$$

The reflection and transmission coefficients in terms of the wave impedances are [2]

$$R = \frac{Z_m - Z_0}{Z_m + Z_0} \qquad (2.42)$$

$$T = \frac{2 Z_m}{Z_m + Z_o}. \qquad (2.43)$$

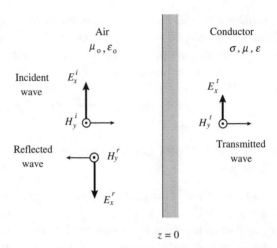

Figure 2.8 Plane wave incident on conducting half space.

Since metal is a much better conductor than air or free space, $Z_m << Z_o$. Then, to a good approximation [1]

$$R \cong -1 \tag{2.44}$$

and

$$T \cong \frac{2Z_m}{Z_o}. \tag{2.45}$$

The reflected and transmitted electric and magnetic fields, from (2.39), (2.40), (2.41), (2.44), and (2.45), are

$$E^r = -E^i \qquad E^t = \frac{2Z_m}{Z_0} E^i$$
$$H^r = H^i \qquad H^t = 2H^i. \tag{2.46}$$

In summary, at the surface of a metal with finite conductivity, to a good approximation, the reflected E field is equal in magnitude and opposite in phase to the incident E field. The transmitted E field is very much smaller than the incident E field (by the ratio $2Z_m/Z_o$). The reflected H field is equal in magnitude and has the same phase as the incident H field. The transmitted H field has twice the magnitude of the incident H field.

An expression for the E field transmission coefficient T in terms of the conductivity and permeability of the metal will now be derived. The intrinsic impedance of free space is

$$Z_o = \sqrt{\frac{\mu_o}{\varepsilon_o}} = 120\pi. \tag{2.47}$$

For a plane wave or TEM wave, the wave impedance is the same as the intrinsic impedance of the medium.

The intrinsic impedance of metal is [1]

$$Z_m = \sqrt{\frac{j\omega\mu}{\sigma}} = (1+j)\sqrt{\frac{\omega\mu}{2\sigma}} \tag{2.48}$$

where $\omega = 2\pi f$ radian frequency, rad/sec
        f frequency, Hz.

The magnitude is

$$|Z_m| = \sqrt{\frac{\omega\mu}{\sigma}} = \sqrt{\frac{2\pi f \mu}{\sigma}}. \tag{2.49}$$

Finally, the E-field transmission coefficient in terms of the conductivity and permeability of the metal is found by substituting (2.47) and (2.48) into (2.45):

$$T = (1+j)\sqrt{\frac{f\mu}{3600\pi\sigma}} \tag{2.50}$$

and

$$|T| = \left|\frac{E^t}{E^i}\right| = \sqrt{\frac{f\mu}{1800\pi\sigma}}. \tag{2.51}$$

The relative conductivities and permeabilities of some common metals are listed in Table 2.1. Since there are many different alloys of iron and steel, the values in the table are only typical. The permeabilities of iron and steel are initial permeabilities (i.e., small impressed field). In addition,

Table 2.1 Relative Conductivities and Permeabilities

Metal	σ_r	μ_r
Aluminum	0.61	1
Brass	0.27	1
Iron	0.18	235 (see note)
Copper	1	1
Steel	0.1	180 (see note)

$\sigma = \sigma_r\sigma_c$ where $\sigma_c = 5.8 \times 10^7$ mhos/m
            (conductivity of copper)

$\mu = \mu_r\mu_o$ where $\mu_o = 4\pi \times 10^{-7}$ henrys/m
            (permeability of free space)

Note: At a frequency of approximately 1 kHz

the permeabilities of iron and steel are a function of frequency: the values listed are for a frequency of approximately 1 kHz. Above 1 MHz, the relative permeability of most metals approaches unity.

The E-field transmission coefficient in dB in terms of σ_r and μ_r is

$$T(dB) = 20 \log |T| = -174 + 10 \log \frac{f \, \mu_r}{\sigma_r}. \tag{2.52}$$

Equation (2.52) is plotted in Fig. 2.9 for copper and steel. The reciprocal of the transmission coefficient (or the negative of the transmission coefficient in dB) is called *reflection loss*.

While the transmitted electric field in metals is orders of magnitude smaller than the incident electric field as evident in Fig. 2.9 (e.g., 114 dB down in copper at 1 MHz), the transmitted magnetic field is twice the magnitude of the incident magnetic field as indicated in (2.46).

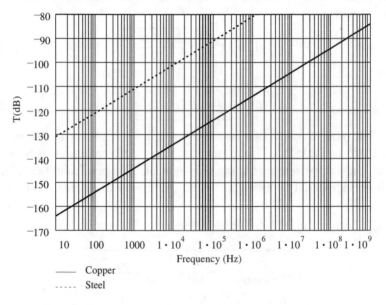

Figure 2.9 E-field transmission coefficient for copper and steel.

2.7   DIFFUSION AND SKIN DEPTH

The transmitted wave in Fig. 2.8 attenuates exponentially in the conductor. The diffusion equation is

$$E_x^t(z) = E_x^t(0)\varepsilon^{-\gamma z} \tag{2.53a}$$

$$H_y^t(z) = H_y^t(0)\varepsilon^{-\gamma z} \tag{2.53b}$$

where $E_x^t(0)$, $H_y^t(0)$ transmitted fields at the surface ($z = 0$)
         $\gamma = \alpha + j\beta$ propagation constant
         α attenuation constant, nepers/m
         β phase constant, rad/m.

The propagation constant γ is

$$\gamma = \sqrt{j\omega\mu\sigma} = \sqrt{\pi f \mu \sigma} + j\sqrt{\pi f \mu \sigma}. \tag{2.54}$$

Then (2.53) can be written as

$$E_x^t(z) = E_x^t(0)\varepsilon^{-\sqrt{\pi f \mu \sigma}\,z}\varepsilon^{-j\sqrt{\pi f \mu \sigma}\,z} \tag{2.55a}$$

$$H_y^t(z) = H_y^t(0)\varepsilon^{-\sqrt{\pi f \mu \sigma}\,z}\varepsilon^{-j\sqrt{\pi f \mu \sigma}\,z}. \tag{2.55b}$$

Skin depth or *depth of penetration* is defined as the distance the wave must travel in order to decay by an amount equal to $\varepsilon^{-1} = 0.368$, or 8.686 dB. From (2.55), the skin depth δ is

$$\delta = \frac{1}{\sqrt{\pi f \mu \sigma}}. \tag{2.56}$$

In a distance equal to five skin depths, the fields are reduced by a factor of 0.0067, or 43.43 dB. For copper, the skin depth is 8.5 mm at 60 Hz and 0.066 mm at 1 MHz.

From (2.55), the attenuation of the fields as a function of distance is

$$A = \frac{\left|E_x^t(z)\right|}{\left|E_x^t(0)\right|} = \frac{\left|H_y^t(z)\right|}{\left|H_y^t(0)\right|} = \varepsilon^{-\sqrt{\pi f \mu \sigma}\,z} \tag{2.57}$$

where

$$\sqrt{\pi f \mu \sigma} = \text{attenuation constant in nepers per meter.}$$

The attenuation in dB is

$$A(\text{dB}) = 20 \log A = -8.686\sqrt{\pi f \mu \sigma}\,z \qquad \text{dB.} \tag{2.58}$$

The reciprocal of the attenuation (or the negative of the attenuation in dB) is called *absorption loss*.

Expressing (2.58) in terms of the relative permeability and conductivity and the distance z in millimeters, we have

$$A(\text{dB}) = -0.1314 z_{mm}\sqrt{f \mu_r \sigma_r} \qquad \text{dB.} \tag{2.59}$$

The attenuation in dB per millimeter of distance traveled is

$$A(\text{dB/mm}) = -0.1314\sqrt{f \mu_r \sigma_r} \qquad \text{dB/mm.} \tag{2.60}$$

Equation (2.60), which applies to both the electric and magnetic fields, is plotted in Fig. 2.10 for copper and steel. Note that below 1 kHz, the

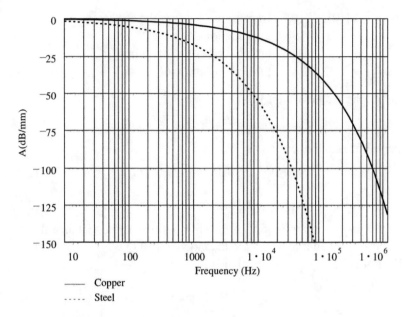

Figure 2.10   Attenuation in dB per mm for copper and steel.

attenuation in copper is less than 4 dB/mm. Above 1 MHz, the attenuation
in copper exceeds 130 dB/mm. The attenuation in steel is significantly
greater than that in copper.

2.8   TRANSMISSION THROUGH A METAL SHEET (SHIELDING EFFECTIVENESS)

Figure 2.11 shows a plane wave incident on a metal sheet of infinite extent
and of thickness τ. We wish to determine how much of the incident wave
penetrates the sheet. The shielding factor is defined as

$$S = \frac{\left|E_2^t\right|}{\left|E^i\right|}. \tag{2.61}$$

In Section 2.6, the transmitted (refracted) E and H fields at the surface
$z = 0$ were derived. In Section 2.7, the attenuation (diffusion) of the E
and H fields in the metal as a function of the distance z was derived. The
only remaining step to determine the shielding factor S is to calculate the
transmission coefficient at the metal-air boundary at $z = \tau$. The reflected
fields at $z = \tau$ and the re-reflected fields at $z = 0$ are neglected because

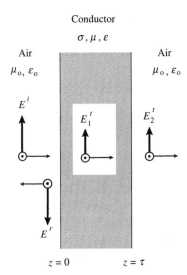

Figure 2.11 Plane wave incident on a metal sheet.

the attenuation in the metal is so high (except at the very low frequencies). See Fig. 2.10.

Let T_2 denote the E-field transmission coefficient at the metal-air boundary at $z = \tau$. By analogy with (2.43)

$$T_2 = \frac{2 Z_o}{Z_o + Z_m}. \tag{2.62}$$

Since $Z_m << Z_o$, to a good approximation

$$T_2 = 2 \tag{2.63}$$

and

$$T_2(dB) = 20 \log |T_2| = 6 \text{ dB}. \tag{2.64}$$

(This is the opposite of the situation at the air-metal interface at $z = 0$, where the transmitted magnetic field was twice the incident magnetic field.)

The shielding factor is then

$$S = \frac{\left| E_2^t \right|}{\left| E^i \right|} = T \, T_2 \, A \tag{2.65}$$

or in dB,

$$S \,(\text{dB}) = T \,(\text{dB}) + T_2(\text{dB}) + A(\text{dB}). \tag{2.66}$$

Finally, substituting (2.52), (2.64), and (2.59) into (2.66)

$$\mathcal{S}\,(\text{dB}) = -168 + 10\log\frac{f\mu_r}{\sigma_r} - 0.1314\,\tau_{mm}\sqrt{f\mu_r\sigma_r} \qquad (2.67)$$

where τ_{mm} thickness of metal sheet, millimeters.

Shielding effectiveness is the reciprocal of the shielding factor $\mathcal{S}$ or, equivalently, the negative of $\mathcal{S}$(dB).

Figures 2.9 and 2.10 can be used to find the shielding factor for copper and steel. First, add +6 dB to the curves in Fig. 2.9 to account for T_2, the transmission coefficient at the surface at $z = \tau$. Next, *multiply* the values in Fig. 2.10 by the thickness of the shield in millimeters. Add the two results. The shielding factor applies to both the electric and magnetic field components of a plane wave. For a copper shield 1 mm thick, the shielding factor is -160 dB at 10 Hz and -240 dB at 1 MHz.

As a practical matter, a thin metal sheet is an effective shield for plane waves, even at very low frequencies. Leakage through apertures is of more importance, especially at higher frequencies. For a more complete discussion of shielding, including near fields, multiple reflections in thin shields, and aperture coupling; see, for instance, Ott [3].

2.9   FIELDS FROM A SHORT DIPOLE

Short dipoles and small loops are the two canonical forms for all electrically small antennas. That is, all electrically small antennas can be represented as a short dipole, small loop, or a combination of these two. See Hansen [4].

The simplest radiator is the short dipole antenna, shown in Fig. 2.12 in spherical coordinates. In many applications, the fields radiated by currents flowing on simple wires and cables can be found by modeling them as a short dipole (or a short monopole). The fields from more complex wire structures can be found by the superposition of the fields of a series-connected array of short dipoles.

The study of the fields from short dipoles and small loops (Section 2.10) also provides insight into the behavior of the fields in the reactive near-field and far-field regions of space surrounding these sources.

The length L of the dipole in Fig. 2.12 is short compared to a wavelength (generally taken to be $L < \lambda/10$, or shorter), and the diameter of the wire is much less than the length.

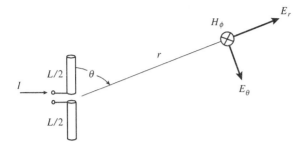

Figure 2.12 Short dipole geometry.

The field quantities in (2.68) to (2.70) are based on the assumption that the current I is uniform (constant) over the length of the dipole. While this current distribution is not physically realizable, it is useful when using series-connected dipoles to approximate actual current distributions on complex wire structures.

In the case of a single short dipole, a triangular current distribution is more representative of the boundary conditions, and thus more accurate. For the triangular distribution, the current at the midpoint is I and the current at the ends of the dipole arms is zero. The fields for the triangular distribution are one-half those for the uniform distribution in (2.68) to (2.70). A thorough discussion of current distributions on wire antennas is given by Balanis [5].

The fields from a short dipole with a uniform current distribution are given below. These equations are from Kraus [6], but in a slightly modified form.

$$E_\theta = 30\,I\,L\,\beta^2\,\sin\theta\left[\frac{j}{\beta r} + \frac{1}{(\beta r)^2} - \frac{j}{(\beta r)^3}\right]\varepsilon^{j\omega t}\varepsilon^{-j\beta r} \qquad (2.68)$$

$$E_r = 60\,I\,L\,\beta^2\,\cos\theta\left[\frac{1}{(\beta r)^2} - \frac{j}{(\beta r)^3}\right]\varepsilon^{j\omega t}\varepsilon^{-j\beta r} \qquad (2.69)$$

$$H_\phi = \frac{I\,L\,\beta^2}{4\pi}\,\sin\theta\left[\frac{j}{\beta r} + \frac{1}{(\beta r)^2}\right]\varepsilon^{j\omega t}\varepsilon^{-j\beta r} \qquad (2.70)$$

where E_θ transverse electric field, V/m
$\quad\quad\quad H_\phi$ transverse magnetic field, A/m
$\quad\quad\quad E_r$ radial electric field, V/m
$\quad\quad\quad r$ distance to field point, m
$\quad\quad\quad I$ current, A

L length of dipole, m

$\beta = \omega/c = 2\pi/\lambda$ phase constant, radians /m

λ wavelength, m

$\omega = 2\pi f$ radians/sec

f frequency, Hz.

Figure 2.13 is a plot, in dB, of E_θ and H_ϕ normalized to $\beta r = 2\pi r/\lambda = 1$, that is,

$$20\log\frac{E_\theta(\beta r)}{E_\theta(\beta r = 1)} \quad \text{and} \quad 20\log\frac{H_\phi(\beta r)}{H_\phi(\beta r = 1)}. \tag{2.71}$$

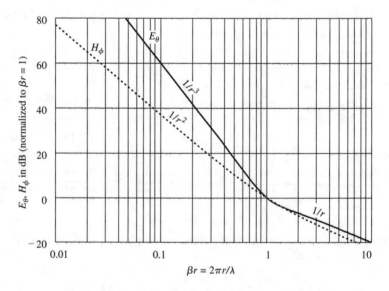

Figure 2.13 Near and far fields of a short dipole.

The transition between the reactive near-field and the far-field regions occurs at a distance of $\beta r = 2\pi r/\lambda = 1$. In the reactive near-field region, E_θ falls off as $1/r^3$ (60 dB per decade distance) and H_ϕ falls off as $1/r^2$ (40 dB per decade distance). In the far-field region, both E_θ and H_ϕ fall off as $1/r$ (20 dB per decade distance) and constitute a propagating plane wave. The radial electric field component E_r, in (2.69), has only $1/r^2$ and $1/r^3$ terms and does not propagate.

In Fig. 2.14, the magnitude of E_θ versus distance r is plotted for three frequencies: 100 kHz, 1 MHz, and 10 MHz. The length of the short dipole is $L = 1$ m, and the current is $I = 1$ ampere. A triangular current distribution is assumed.

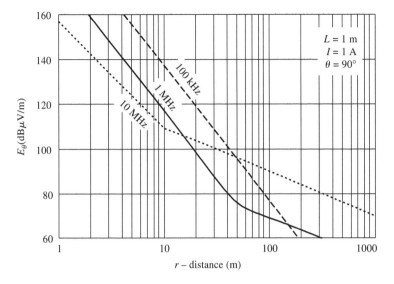

Figure 2.14 Electric field strength versus distance for a short dipole.

Note that in the reactive near-field region, the field strength is greater at the lower frequencies. Conversely, in the far-field region, the field strength increases as the frequency increases.

2.10   FIELDS FROM A SMALL LOOP

Small loops are generally not used as transmitting antennas because they have small radiation resistances compared to short dipoles. However, many unintentional sources of radiation are essentially loop antennas. Examples include transformers, inductors, two-wire transmission lines, and printed circuit boards.

Small loops come in many different shapes, the most common being square and circular. The shape is of no consequence. It is the loop area S that determines the magnitude of the radiated fields.

A small circular loop antenna is shown in Fig. 2.15. The circumference of the loop is small compared to a wavelength and the current I is uniform in amplitude and phase. The fields from a small loop are given below. These equations are a slightly modified form of those due to Schelkunoff and Friis [7].

$$E_\phi = -j \, 30 \, I \, S \, \beta^3 \, \sin \theta \left[\frac{j}{\beta r} + \frac{1}{(\beta r)^2} \right] \varepsilon^{j\omega t} \varepsilon^{-j\beta r} \qquad (2.72)$$

$$H_\theta = \frac{j\,I\,S\,\beta^3}{4\pi}\,\sin\theta\left[\frac{j}{\beta r} + \frac{1}{(\beta r)^2} - \frac{j}{(\beta r)^3}\right]\varepsilon^{j\,\omega\,t}\varepsilon^{-j\beta r} \qquad (2.73)$$

$$H_r = \frac{j\,I\,S\,\beta^3}{2\pi}\,\cos\theta\left[\frac{1}{(\beta r)^2} - \frac{j}{(\beta r)^3}\right]\varepsilon^{j\omega t}\varepsilon^{-j\beta r} \qquad (2.74)$$

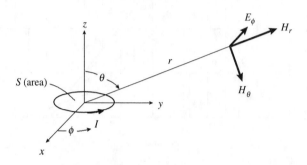

Figure 2.15 Small circular loop geometry.

where E_ϕ transverse electric field, V/m
$\quad\;\;\;\,H_\theta$ transverse magnetic field, A/m
$\quad\;\;\;\,H_r$ radial magnetic field, A/m
$\quad\;\;\;\,r$ distance to field point, m
$\quad\;\;\;\,I$ current, A
$\quad\;\;\;\,S$ area of loop, m^2
$\quad\;\;\;\,\beta$ $= \omega/c = 2\pi/\lambda$ phase constant, rad/m
$\quad\;\;\;\,\lambda$ wavelength, m
$\quad\;\;\;\,\omega$ $= 2\pi f$ rad/sec
$\quad\;\;\;\,f$ frequency, Hz.

    In the reactive near-field region ($\beta r < 1$) the transverse magnetic field component H_θ falls off as $1/r^3$ (60 dB per decade distance) and the transverse electric field component E_ϕ falls off as $1/r^2$ (40 dB per decade distance). Compare this with the transverse field components for a short dipole in Fig. 2.13, where the transverse *electric* field falls off as $1/r^3$ and the transverse *magnetic* field falls off as $1/r^2$. The curves in Fig. 2.13 for a short dipole are identical to those for a small loop if E_θ and H_ϕ are replaced by H_θ and E_ϕ, respectively.

    As with the short dipole, in the far-field region of a small loop ($\beta r > 1$), both transverse field components E_ϕ and H_θ fall off as $1/r$ (20 dB per decade distance) and constitute a propagating plane wave. The radial magnetic field component H_r, (2.74), has only $1/r^2$ and $1/r^3$ terms and does not propagate.

Measurements of the rate of attenuation (fall-off) of the transverse field components in the reactive near-field region can be used to determine if a source of radiation is an electric dipole or a magnetic loop. For example, suppose a receiving loop antenna is used to measure the transverse *magnetic* field component at two or more distances from the source, and the data is plotted (preferably in dBμA/m versus distance on a logarithmic scale similar to Fig. 2.13). If the measured fall-off is $1/r^3$ (60 dB per decade distance or 18 dB per octave distance), the source is a magnetic loop. See (2.73). If the fall-off is $1/r^2$ (40 dB per decade distance or 12 dB per octave distance), the source is an electric dipole. See Fig. 2.13 and (2.70).

Conversely, suppose a dipole receiving antenna is used to measure the transverse *electric* field component at two or more distances from the source, and the data is plotted (again, preferably in dBμV/m versus distance on a logarithmic scale similar to Fig. 2.13). If the measured fall-off is $1/r^3$ (60 dB per decade distance or 18 dB per octave distance), the source is an electric dipole. See Fig. 2.13 and (2.68). If the fall-off is $1/r^2$ (40 dB per decade distance or 12 dB per octave distance), the source is magnetic loop. See (2.72).

If the measured fall-off of either the transverse electric or transverse magnetic field component is $1/r$ (20 dB per decade distance or 6 dB per octave distance), then the measurements are in the far-field and it cannot be determined whether the source is an electric dipole or a magnetic loop. It is assumed that the ground-reflected wave is negligible in all of the above.

2.11   NEAR-FIELD AND FAR-FIELD REGIONS

The space surrounding an antenna or other source of radiation is divided into three regions: the reactive near-field region, the radiating near-field region and the far-field region. These are depicted in Fig. 2.16. The characteristics of the fields in these regions are discussed below. See [8] for detailed definitions of these field regions.

In the *reactive near-field* region,

- the reactive fields predominate. These fields fall off as $E \propto 1/r^n$ and $H \propto 1/r^m$ where n, m > 1.
- electromagnetic energy is stored in the fields.

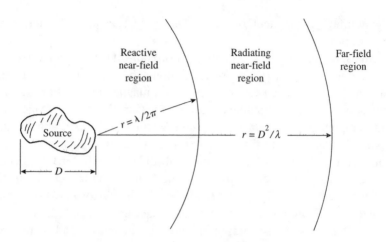

Figure 2.16 Field regions.

- a large radial field component exists.
- the outer boundary of the reactive near-field region is $r = \lambda/2\pi$ where λ is the wavelength.

The *radiating near-field* region has the following characteristics:

- radiation fields predominate
- prediction of the variation of the fields with distance is not possible because the fields oscillate out to a distance equal to $D^2/4\lambda$. See Adams et al. [9].
- this region lies between the reactive near-field region and the far-field region.
- the angular field distribution (the field pattern) depends on the distance.
- this region may not exist if $D << \lambda$ where D is the maximum overall dimension of the source.
- sometimes referred to as the Fresnel region by analogy to optical terminology.

The *far-field* region has the following characteristics:

- the electric and magnetic field components are transverse to the direction of propagation and to each other. That is, they constitute a plane wave.
- the electric and magnetic fields fall off as $1/r$ (inverse distance).

- the angular field distribution (the field pattern) is independent of distance.

- the transition between the radiating near-field region and the far-field region is commonly taken as $r = D^2/\lambda$ (see Hansen [10]) but this distance may vary between $D^2/2\lambda$ and $2D^2/\lambda$ depending on how much deviation from a $1/r$ fall-off can be tolerated in a particular application.

- it is sometimes referred to as the Fraunhofer region by analogy to optical terminology.

Two examples will be used to illustrate the practical application of these concepts. The first example is a microwave oven, shown in Fig. 2.17. The maximum overall dimension of the oven is $D = 0.5$ m, and the oven radiates at a frequency of 2450 MHz. (The source could be any device or antenna having the same size and radiating at the same frequency.) The reactive near-field region only extends out to a distance of $\lambda/2\pi = 2$ cm at this short wavelength. The transition between the radiating near-field region and the far-field region takes place at a distance of $D^2/\lambda = 2.1$ m.

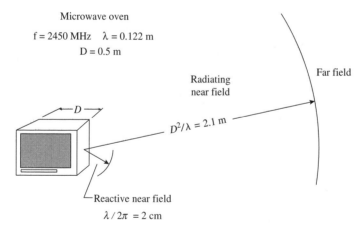

Figure 2.17   Field regions around a microwave oven.

The second example is a personal computer system, shown in Fig. 2.18. The maximum overall dimension of the PC system is the same as that of the microwave oven. The emission from the PC, however, is at a much lower frequency: $f = 1.5$ MHz from the switching power supply. In this example, the reactive near-field region extends out to a distance of 31.8 m. The radiating near-field region does not exist because the dimensions of the source are much smaller than a wavelength ($D^2/\lambda = 1.25$ mm).

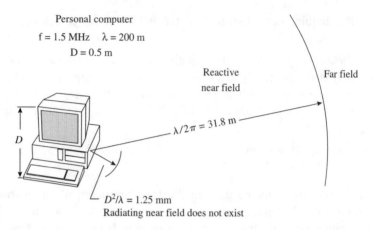

Personal computer
f = 1.5 MHz   λ = 200 m
         D = 0.5 m

Reactive   Far field
near field

$\lambda/2\pi = 31.8$ m

D

$D^2/\lambda = 1.25$ mm
Radiating near field does not exist

Figure 2.18 Field regions around a personal computer system.

2.12 WAVE IMPEDANCE

The wave impedance of an electromagnetic wave is defined as the ratio of the transverse electric field component to the mutually perpendicular transverse magnetic field component [2]:

$$Z_w = E_t/H_t. \tag{2.75}$$

The modifier transverse means that E_t and H_t lie in a plane that is perpendicular to the direction of propagation, as illustrated in Fig. 2.19.

The definition of wave impedance applies to TEM waves (transverse electromagnetic waves) and to waves having longitudinal components, for example, TE (transverse electric) and TM (transverse magnetic) waves. Included in these categories are waves in free space and other unbounded media, and guided waves on transmission lines and in waveguides.

The wave impedance of a **plane wave or TEM wave** is equal to the intrinsic impedance of the medium in which the wave propagates and is

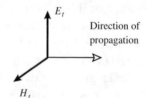

E_t

Direction of
propagation

Figure 2.19 Plane wave. H_t

given by

$$Z_w = \sqrt{\frac{\mu}{\varepsilon}} \tag{2.76}$$

where μ and ε are the permeability and permittivity of the medium, respectively.

The wave impedance of a **plane wave in free space** is denoted as Z_o and is equal to

$$Z_o = \sqrt{\frac{\mu_o}{\varepsilon_o}} = 120\,\pi = 377 \text{ ohms.} \tag{2.77}$$

The wave impedance of the fields from a **short dipole** radiator, from (2.68) and (2.69), is

$$Z_w = \begin{cases} -j\,120\,\pi(1/\beta r) & \text{near field} \quad (\beta r < 1) \\ 120\,\pi & \text{far field} \quad\;\; (\beta r \geq 1). \end{cases} \tag{2.78}$$

The near-field wave impedance of a short dipole is capacitive and greater than $120\,\pi$ ohms. The closer to the dipole in terms of wavelength, the higher the wave impedance. For example, at $\beta r = 2\pi r/\lambda = 0.01$, the wave impedance is 37,700 ohms.

Wave impedance, defined as the ratio of the transverse E- and H-field components, conveys no information about the longitudinal field. In the near-field region of a dipole radiator, the radial E field is twice the magnitude of the transverse E field and might be the predominant component in coupling and interference situations.

The wave impedance of the fields from a **small loop** antenna, from (2.72) and (2.73), is

$$Z_w = \begin{cases} j\,120\,\pi\;\beta\,r & \text{near field} \quad (\beta\,r < 1) \\ 120\,\pi & \text{far field} \quad\;\; (\beta\,r \geq 1). \end{cases} \tag{2.79}$$

The near-field wave impedance of a small loop antenna is inductive and less than $120\,\pi$ ohms. The closer to the loop in terms of wavelength, the lower the wave impedance. For example, at $\beta r = 2\pi r/\lambda = 0.01$, the wave impedance is 3.77 ohms.

In the near-field region of a loop antenna, the radial H field is twice the magnitude of the transverse H field and might be the predominant component in coupling and interference situations.

2.13   POWER DENSITY AND HAZARDOUS RADIATION

Human exposure to high-level electromagnetic fields can cause harmful effects. Recommended levels of maximum permissible exposure (MPE) to radio frequency electromagnetic fields are given in IEEE C95.1-1991 [11]. The MPEs are given in terms of the rms electric (E) and magnetic (H) field strengths, and in terms of the equivalent free-space plane-wave average power density S_{AV}. The relationships between field strengths and power density in the reactive near-field and far-field regions are reviewed below.

The instantaneous magnitude and direction of power flow per unit area (power density) in an electromagnetic field are given by the instantaneous Poynting vector **S**:

$$\mathbf{S} = \mathbf{E} \times \mathbf{H} \qquad \text{watts/m}^2. \tag{2.80}$$

For a sinusoidal time varying field, the average Poynting vector is

$$\mathbf{S}_{AV} = \frac{1}{2} \operatorname{Re} \mathbf{E} \times \mathbf{H}^* = \operatorname{Re} \mathbf{E}_{\text{rms}} \times \mathbf{H}^*_{\text{rms}} \tag{2.81}$$

where $\mathbf{H}^*$ denotes the complex conjugate of $\mathbf{H}$ and where the *rms* values of $\mathbf{E}$ and $\mathbf{H}$ are equal to the peak values divided by $\sqrt{2}$.

Figure 2.20 illustrates the electric and magnetic fields as space vectors and as phasors in the time domain. The space angle between the field vectors is α, and the phase angle between E and H is θ.

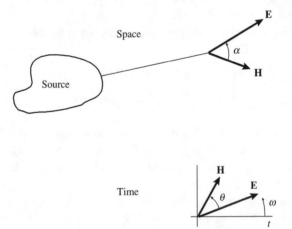

Figure 2.20 Space and time-domain representation of E and H fields.

At a single frequency, the magnitude of the average power density is

$$S_{AV} = E_{\text{rms}} H_{\text{rms}} \sin\alpha \cos\theta \qquad \text{W/m}^2. \qquad (2.82)$$

Equation (2.82) is general and applies to both reactive near fields and plane waves.

There is no commercial instrument which can measure the true average power-density expressed in (2.82) in the reactive near-field region. Commercial radiation hazard meters are calibrated in units of *equivalent* free-space plane-wave average power density, but use probes which measure either the electric field component or the magnetic field component. Typical E-field probes use two or three orthogonal dipoles. Typical H-field probes use three orthogonal loops.

Care should be taken when interpreting readings taken with radiation hazard meters in the reactive near field. In high-impedance fields, measurements with H-field probes will read lower than the true power density, in some cases by orders of magnitude, depending on the wave impedance. Conversely, in low-impedance fields, measurements with E-field probes will read lower than the true power density. This is the reason that the maximum permissible exposures in IEEE C95.1 are stated in terms of the rms electric (E) and magnetic (H) field strengths, and in terms of the *equivalent* free-space plane-wave average power density S_{AV}.

For a plane wave in free space with sinusoidal time variation,

$$E = 120\,\pi\ H$$

$$\sin\alpha = 1$$

$$\cos\theta = 1$$

and (2.82) can be expressed as

$$S_{AV} = \frac{E_{\text{rms}}^2}{3770} \qquad \text{mW/cm}^2 \qquad (2.83)$$

or

$$S_{AV} = 37.7\ H_{\text{rms}}^2 \qquad \text{mW/cm}^2 \qquad (2.84)$$

or

$$S_{AV} = \frac{E_{\text{rms}} H_{\text{rms}}}{10} \qquad \text{mW/cm}^2 \qquad (2.85)$$

where E is in V/m, H is in A/m, and S_{AV} is in milliwatts per square centimeter.

The maximum permissible exposures for an uncontrolled environment from IEEE C95.1-1991 are shown in Table 2.2. E, H, and power density as a function of frequency are plotted in Fig. 2.21.

Table 2.2 Maximum Permissible Exposure for Uncontrolled Environments

Frequency Range (MHz)	E (V/m)	H (A/m)	Power Density (mW/cm²)
0.003–0.1	614	163	
0.1–1.34	614	16.3/f	
1.34–3.0	823.8/f	16.3/f	
3.0–30	823.8/f	16.3/f	
30–100	27.5	$158.3/f^{1.668}$	
100–300	27.5	0.0729	0.2
300–3000			f/1500
3000–15,000			f/1500
15,000–300,000			10.0

Note: f = frequency in MHz.

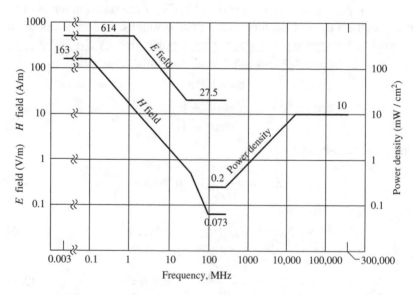

Figure 2.21 Maximum permissible exposure for an uncontrolled environment. (IEEE C95.1-1991). *Copyright ©1992, IEEE. All rights reserved.*

REFERENCES

[1] J. D. Kraus and K. R. Carver, *Electromagnetics*, second edition, McGraw-Hill, New York, 1973.

[2] R. E. Collin, *Field Theory of Guided Waves*, second edition, IEEE Press, New York, 1991.

[3] H. W. Ott, *Noise Reduction Techniques in Electronic Systems*, John Wiley & Sons, New York, 1976.

[4] R. C. Hansen, *Reference Data for Radio Engineers*, Chapter 32 - Antennas, Howard W. Sams & Co., Division of MacMillan, Inc., Indianapolis, Indiana, seventh edition, 1989.

[5] C. A. Balanis, *Antenna Theory: Analysis and Design*, Harper and Row, Publishers, Inc., New York, 1982.

[6] J. D. Kraus, *Antennas*, McGraw-Hill, New York, 1950.

[7] S. A. Schelkunoff and H. T. Friis, *Antennas: Theory and Practice*, John Wiley & Sons, New York, 1952.

[8] IEEE Std 100-1977, *IEEE Standard Dictionary of Electrical and Electronics Terms*, published by IEEE, New York, distributed in cooperation with Wiley-Interscience, New York, 1977.

[9] A. T. Adams, Y. Leviatan and K. S. Nordby, "Electromagnetic Near Fields as a Function of Electrical Size," *IEEE Transactions on Electromagnetic Compatibility*, vol. EMC-25, no. 4, pp. 428–432, November 1983.

[10] R. C. Hansen and L. L. Bailin, "A New Method of Near Field Analysis," *IRE Transactions on Antennas and Propagation*, pp. S458–S467, December 1959.

[11] IEEE C95.1-1991, *IEEE Standard for Safety Levels with Respect to Human Exposure to Radio Frequency Electromagnetic Fields, 3 kHz to 300 GHz*, Institute of Electrical and Electronic Engineers, 1992. Recognized as an American National Standard (ANSI).

CHAPTER 3

PROPAGATION

Radio wave propagation is defined as the transfer of energy by electromagnetic radiation at radio frequencies [1]. Radio waves can propagate in a variety of modes depending on the path geometry, the frequency, and the electrical properties and temporal variability of the earth's surface, the atmosphere, the troposphere and the ionosphere. The most common propagation processes or modes are summarized in Table 3.1. Ionospheric, tropospheric, and earth-space propagation are of limited general interest and are not addressed here.

This chapter begins with a review of the least complicated mode of propagation—free space. Only the direct wave is of consequence in free-space propagation. Reflections, diffraction, refraction and absorption are negligible. Applications include propagation in anechoic chambers, on elevated antenna ranges, and in regions close to a source where reflections are negligible compared with the direct wave.

The next topic reviewed in this chapter is ground-wave propagation over plane earth. This is followed by an examination of the limiting case of propagation over a perfectly conducting ground plane ($\sigma = \infty$). The final subjects covered in this chapter are the attenuation of electromagnetic fields by building structures, edge diffraction, and the Rayleigh roughness criterion.

3.1 FREE SPACE PROPAGATION

Consider an isotropic radiator in free space as shown in Fig. 3.1. By definition, an isotropic radiator has equal radiation intensity in all directions. In free space, the resultant spherical wave spreads out with uniform intensity in all directions. If P_o is the power radiated by the isotropic source, the

TAble 3.1 Radio Wave Propagation Modes

Mode	Characteristics
Free space	Direct wave only (no reflections, diffraction, refraction or absorption) Examples: • Regions near source • Anechoic chambers • Elevated antenna ranges
Ground wave	Plane earth (line-of-sight) region • Space wave and surface wave • Distances to 50 miles at 1 MHz, 5 miles at 1 GHz Spherical earth (diffraction) region: • Diffraction theory (ray theory inadequate) • Distances to 2000 miles at VLF, 200 miles at HF
Sky wave	Sky wave reflected from ionosphere 3 to 30 MHz region of spectrum (HF) D, E, F_1, and F_2 layers Height of ionosphere—50–40 km Distances of 100 to many thousands of miles
Tropospheric scatter	Forward scatter and backscatter Height of troposphere—10km VHF to SHF bands Distances to hundreds of miles beyond the horizon
Earth-space	Satellite communications—1 to 10 GHz Limited by cosmic and solar noise and by refraction and absorption in the troposphere and ionosphere

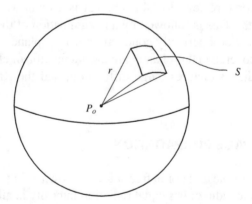

Figure 3.1 Isotropic radiator in free space.

power density S at a radial distance r is just the radiated power divided by the surface area of the surrounding imaginary sphere:

$$S = \frac{P_o}{4\pi r^2} \qquad \text{W/m}^2. \tag{3.1}$$

In the far field, power density S and electric field strength E are related by

$$S = E^2/120\pi \tag{3.2}$$

where 120π is the intrinsic impedance of free space.

The electric field strength at a radial distance r is

$$E_o = \frac{\sqrt{30P_o}}{r}\varepsilon^{-j\beta r} \qquad \text{V/m} \tag{3.3}$$

where the magnitude follows from (3.1) and (3.2) and where $\beta = 2\pi/\lambda$ is the free-space wave number or phase constant. (β is the space rate of decrease of the phase of the field in the direction of propagation.) The subscript on E_o denotes the free space field.

Note that the power density S falls off as $1/r^2$ while the electric field strength E_o falls off as $1/r$.

Now consider a source with a directive gain pattern $D(\theta, \phi)$ shown in Fig. 3.2. In this case, the radiated power is concentrated in some direction (or directions) as it spreads out over the surface of a surrounding imaginary sphere. (At sufficient distances from the source, the spherical wave is well approximated by a plane wave.)

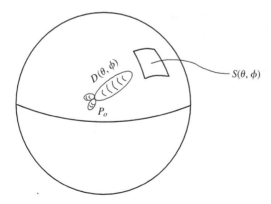

Figure 3.2 Radiator with directive gain pattern.

The far-field power density is

$$S(\theta, \phi) = \frac{P_o D(\theta, \phi)}{4\pi \, r^2} \qquad \text{W/m}^2 \tag{3.4}$$

and the far-field electric field strength is

$$E_o(\theta, \phi) = \frac{\sqrt{30 P_o D(\theta, \phi)}}{r} \varepsilon^{-j\beta r} \qquad \text{V/m.} \tag{3.5}$$

Basic transmission loss L is defined as the path loss between two lossless unity-gain antennas separated by a distance r. In free space, the basic transmission loss is

$$L_{FS} = \frac{P_o}{P_R} = \left[\frac{4\pi r}{\lambda}\right]^2 \tag{3.6}$$

where P_o the power radiated by the transmitting antenna (equal to the input power since the antenna is lossless and has unity gain)

P_R the power received by the receiving antenna.

The free-space basic transmission loss in dB is

$$L_{FS}(dB) = 10 \log L_{FS}$$

or

$$L_{FS}(dB) = 32.44 + 20 \log f_M + 20 \log r_{km} \qquad \text{dB} \tag{3.7}$$

where f_M frequency in megahertz
r_{km} distance in kilometers.

Basic transmission losses for the other radio wave propagation processes in Table 3.1 are much more complicated and are determined by the path geometry, the frequency, the distance, the applicable propagation modes, and the electrical properties and temporal variations of the earth's surface, the troposphere, and the ionosphere.

3.2 GROUND-WAVE PROPAGATION OVER PLANE EARTH

Ground-wave propagation models are partitioned into three regions:

- the co-site region near the source where surface effects are weak or negligible

- the plane earth or line-of-sight region where geometrical optics (ray theory) is valid [2]–[3]

- the spherical earth or diffraction region where geometrical optics breaks down and diffraction theory is required [4]–[6].

These regions are illustrated in Fig. 3.3. In the co-site region, the ground wave attenuates at a rate of essentially $1/r$ (20 dB/decade distance), though peaks and nulls may be present due to the constructive and destructive interference of the ground-reflected wave. This region extends out to a distance of approximately 10 or 20 m as illustrated in Figs. 3.6 and 3.7.

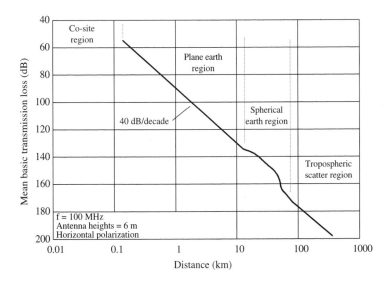

Figure 3.3 An example of the different ground-wave propagation regions. [©*1977 IEEE. Adapted from M.N. Lustgarten and J.A. Madison "An Empirical Propagation Model (EPM-73)", IEEE Transactions on Electromagnetic Compatibility, vol. EMC-19, no. 3, pp. 301–309, August 1977.*]

The plane earth region extends out to a distance of approximately

$$r_{pe} = \frac{80}{\sqrt[3]{f_M}} \quad \text{km} \tag{3.8}$$

where f_M is the frequency in MHz. See Table 3.2 for some examples.

Beyond this distance is the spherical earth (diffraction) region, and beyond that is the region where troposcatter effects become significant.

Table 3.2 Extent of Plane Earth Region

f_M MHz	r_{pe} kilometers	r_{pe} miles
1	80	50
10	37	23
100	17	11
1000	8	5

The geometry for ground wave propagation over plane earth is shown in Fig. 3.4. In the plane earth region, the ground wave is composed of the space wave and the surface wave. The space wave consists of the direct wave and the ground-reflected wave.

GROUND WAVE = <u>DIRECT WAVE + REFLECTED WAVE</u> + SURFACE WAVE

                      SPACE WAVE

In Fig. 3.4,

R_1 distance traveled by direct wave between source and field points
R_2 distance traveled by reflected wave between source and field points
r distance between source and field points measured on the plane
h_1 height of source above the earth plane
h_2 height of field point above the earth plane
θ angle of incidence of reflected wave
γ $= 90° - \theta$ elevation angle, measured from the horizontal
κ relative dielectric constant of the earth referred to free space
σ conductivity of the earth, siemens per meter (S/m).

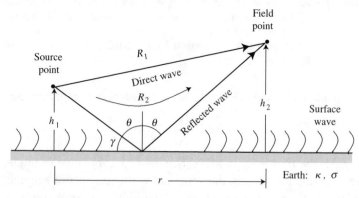

Figure 3.4 Ground-wave geometry in the plane earth region.

Ground-Wave Model

The ground-wave field strength over plane earth from the Burrows-Gray model [4] is

$$E(r) = 2\,E_o(r)\,A(r)\,G(h_1)\,G(h_2) \tag{3.9}$$

where $E(r)$ ground-wave field strength at distance r and height h_2

$E_o(r)$ free space electric field strength at a distance r from the source. $2E_o$ is the field for vertical polarization over a perfectly conducting plane.

$A(r)$ surface wave attenuation factor

$G(h)$ height gain factor (where $h = h_1$ or h_2), independent of distance.

In the Burrows-Gray formulation, $A(r)$, $G(h_1)$, and $G(h_2)$ are functions of frequency, polarization, κ, and σ. Equation (3.9) is valid when $2\pi h_1 h_2/\lambda r < 1$ and is applicable for both space and surface waves. In addition, (3.9) applies only in the far field.

Surface Wave Predominates. The surface wave extends some height above the surface of the earth. This height is a function of frequency and the earth's constants κ and σ. The surface wave is limited to a region of about one wavelength above good earth and five to ten wavelengths above sea water. At greater heights, the space wave is larger in magnitude [7]. (The surface-wave component does not exist above a perfectly conducting plane, i.e., when $\sigma = \infty$.) At heights where the surface wave predominates, the height gain functions are equal to unity: $G(h_1) = 1$ and $G(h_2) = 1$. That is, the ground-wave field strength is not a function of the height of the source (transmit antenna) or the height of the field point (receiving antenna). This is illustrated in Figs. 3.9 and 3.10 at those frequencies and heights where the slope of the height gain curve is zero.

At greater heights where the space wave predominates, the field strength increases with height at a rate proportional to h. That is, $G \propto h$ and the slope of the height gain curve is unity (20 dB per decade height).

When the slope of the height gain curve is $\frac{1}{2}$ (10 dB per decade height), the surface wave and space wave are equal in magnitude.

In the co-site region (the region near the source), the space-wave magnitude exceeds the surface-wave magnitude. In addition, reactive near fields may be significant.

Space Wave Predominates. When the space wave predominates, the method used to calculate the ground-wave field strength depends on the geometry and wavelength.

When the inequality $2\pi h_1 h_2/\lambda r < 1$ is satisfied, the ground-wave field strength can be calculated from either the Burrows-Gray model in (3.9), or directly from ray theory as the sum of the direct and ground-reflected waves. When $2\pi h_1 h_2/\lambda r < 1$, there are no peaks and nulls in the field strength versus distance curves caused by constructive and destructive interference between the direct and ground reflected waves.

When $2\pi h_1 h_2/\lambda r > 1$, the ground-wave field strength must be calculated from ray theory as the sum of the direct wave and the ground-reflected wave.

Refer to Fig. 3.4 and (3.3). The electric field strength at the receiving point due to an isotropic source antenna ($D(\theta, \phi) = 1$) with radiated power P_o is the sum of the direct and ground-reflected waves:

$$E(r) = E_o(R_1) + \rho E_o(R_2) \tag{3.10}$$

or

$$E(r) = \sqrt{30\,P_o}\left[\frac{\varepsilon^{-j\beta R_1}}{R_1} + |\rho|\,\frac{\varepsilon^{-j\beta R_2}\varepsilon^{-j\phi}}{R_2}\right] \tag{3.11}$$

where $\rho = |\rho|\,\varepsilon^{-j\phi}$ reflection coefficient at the surface
$|\rho|$ reflection coefficient magnitude
ϕ reflection coefficient phase
$R_1 = [r^2 + (h_1 - h_2)^2]^{1/2}$ direct-ray path length
$R_2 = [r^2 + (h_1 + h_2)^2]^{1/2}$ reflected-ray path length.

Equation (3.11) is for the far-field region of the source. In the reactive near-field region, the distances R_1 and R_2 in the *denominators* in (3.11) would be squared or cubed, depending on whether the source was a dipole or loop. See (2.68) and (2.72), for example.

Equation (3.11) is valid for horizontal dipoles (receiving and transmitting) which are oriented normal to the direction of propagation, since $D(\theta) = 1$. In general, however, the first and second terms of (3.11) must be modified to account for the pattern directivity of antennas (for example, vertical dipoles and high gain antennas such as microwave horns) especially at large elevation angles, γ.

The reflection coefficient at the surface of a finitely conducting plane earth is a function of frequency, the elevation angle of the incident wave γ, and the electrical constants of the earth.

For horizontal polarization [7]

$$\rho_h = \frac{\sin\gamma - (\kappa - j60\lambda\sigma - \cos^2\gamma)^{1/2}}{\sin\gamma + (\kappa - j\,60\,\lambda\,\sigma - \cos^2\gamma)^{1/2}} \qquad (3.12)$$

and for vertical polarization [7]

$$\rho_v = \frac{(\kappa - j\,60\,\lambda\,\sigma)\,\sin\gamma - (\kappa - j\,60\,\lambda\,\sigma - \cos^2\gamma)^{1/2}}{(\kappa - j\,60\,\lambda\,\sigma)\,\sin\gamma + (\kappa - j\,60\,\lambda\,\sigma - \cos^2\gamma)^{1/2}} \qquad (3.13)$$

where $\kappa = \varepsilon/\varepsilon_o$ is the relative permittivity of the earth
$\quad\;\varepsilon$ permittivity of the earth, F/m
$\quad\;\varepsilon_o$ permittivity of free space, 8.854×10^{-12} F/m
$\quad\;\sigma$ conductivity of the earth, siemens per meter (S/m).

Typical values of conductivity and relative permittivity for different earth terrains are given in Table 3.3. The values for the relative permittivities (dielectric constants) are constant up to 10 GHz. Conductivities are constant up to approximately 100 MHz. Above 100 MHz, the conductivity for different types of ground varies with frequency. See [8].

Table 3.3 Earth Conductivities and Relative Permittivities

Terrain	σ Conductivity (siemens/meter)	κ Relative Permittivity
Sea water	5	80
Fresh water	0.003	80
Good earth	0.01	15
Dry sand	0.001	10

The magnitude and phase of the horizontal and vertical reflection coefficients over good earth ($\kappa = 15$ and $\sigma = 0.01$ S/m) are plotted in Fig. 3.5 for frequencies of 10 MHz and 100 MHz [9]. Note that the magnitude of the vertical reflection coefficient goes through a minimum and the phase goes through $-90°$ at an angle of incidence θ_B known as the pseudo Brewster angle (by analogy with the Brewster angle for perfect dielectrics when $\rho_v = 0$, that is, no reflection takes place). Also note that at grazing incidence ($\gamma = 0$) the phase of the horizontal reflection coefficient is $180°$ and the phase of the vertical reflection coefficient is $-180°$.

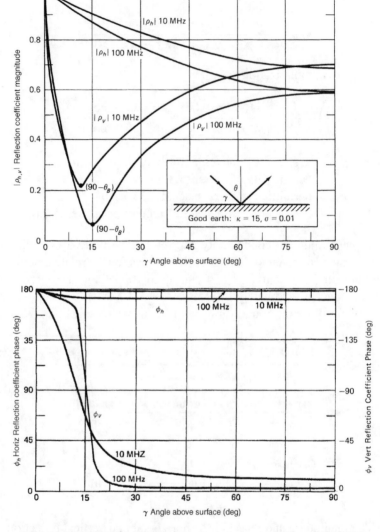

Figure 3.5 Reflection coefficients for good earth at 10 MHz and 100 MHz.

Attenuation and Height Gain Curves

Attenuation and height gain curves for ground-wave propagation over good earth from [10] are shown in Figs. 3.6 to 3.10.[1] Additional curves over dry sand and sea water can be found in [10].

[1] Figures 3.6 to 3.10 ©1969 IEEE. From A. A. Smith, Jr., "Electric Field Propagation in the Proximal Region," *IEEE Transactions on Electromagnetic Compatibility*, vol. EMC-11, no. 4, pp. 151–163, November, 1969.

The electric field attenuation curves in Figs. 3.6 to 3.8 cover frequencies up to 1 GHz and distances to 10 km (6.2 miles) and were calculated using either the Burrows-Gray model or ray theory, as appropriate, for the particular frequency and distance. In the reactive near field, a short dipole source was assumed. (See Section 2.9 for the fields of a short dipole.) For horizontal polarization, and for vertical polarization for frequencies of 30 MHz and greater, $h_1 = h_2 = 1$ m. For vertical polarization for frequencies of 30 MHz and below, the height of the source is $h_1 = 1$ m and the height of the field point is $h_2 = 0$ (representing a monopole receiving antenna on the surface). The quantity in brackets [E′] in Figs. 3.6 to 3.8 is the ratio (in dB) of the field strength at $r = 1$ m in the presence of earth to the field strength that would exist if earth was replaced by free space.

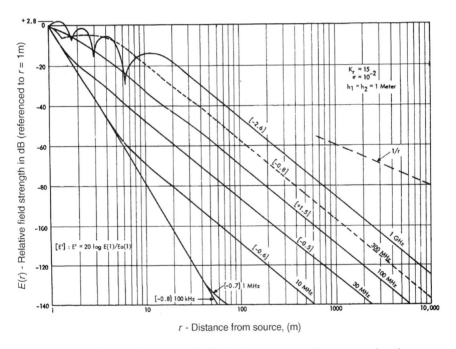

Figure 3.6 Electric field attenuation for horizontal polarization over good earth.
[©*1969 IEEE.*]

Refer to the attenuation curves for horizontal polarization in Fig. 3.6. At frequencies of 10 MHz and greater, the space wave is the principal part of the field. Below 10 MHz, the surface wave predominates except in the region near the source. Note that the initial fall-off rate for the 100 kHz, 1 MHz, and 10 MHz curves is $1/r^4$, which results from a $1/r^3$ factor for the reactive near field and a $1/r$ factor for the surface- or space-wave attenuation. Note also that in the far field, the attenuation rates of the curves in Fig. 3.6 converge to $1/r^2$—a $1/r$ free-space fall-off and a $1/r$ factor for the surface-wave or space-wave attenuation rate.

The attenuation curves for vertical polarization are shown in Figs. 3.7 and 3.8. The surface wave predominates up to approximately 100 MHz for the geometries indicated. Above approximately 100 MHz, the space wave predominates. In the far field, the attenuation rates of the curves in Figs. 3.7 and 3.8 converge to $1/r^2$.

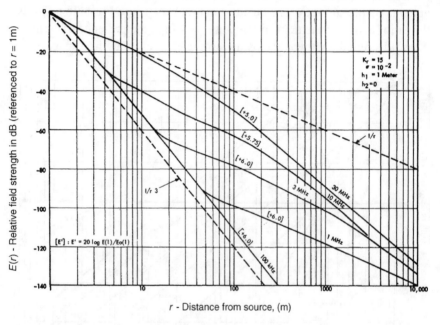

Figure 3.7 Electric field attenuation for vertical polarization over good earth ($f \leq 30$ MHz). [©1969 IEEE.]

The electric field height-gain factors for horizontal and vertical polarizations over good earth are shown in Figs. 3.9 and 3.10. These curves were calculated using the curves of $G(\chi)$ versus $|\chi|$ in the Burrows-Gray paper [4].

The height gain curves $\bar{G}(h)$ in Figs. 3.9 and 3.10 are normalized to a height of $h = 1$ m. The quantities $[G']$ on the curves are the height gains at $h = 1$ m. To find the absolute height gain at any height $h > 1$ m, add $\bar{G}(h)$ and $[G']$.

The height gain curves are valid when $2\pi h_1 h_2 / \lambda r < 1$ and are independent of distance. Also, the curves apply independently to the source

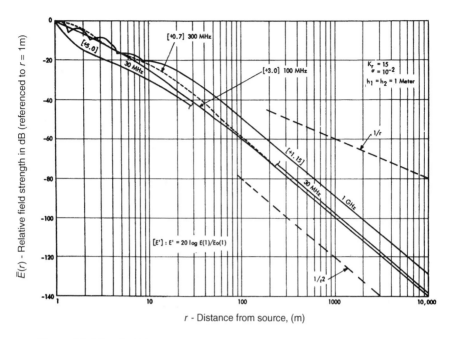

Figure 3.8 Electric field attenuation for vertical polarization over good earth (f > 30 MHz). [©*1969 IEEE*.]

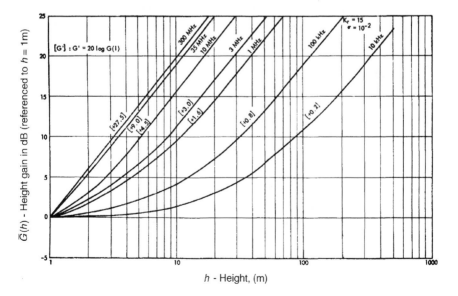

Figure 3.9 Height gain for horizontal polarization over good earth. [©*1969 IEEE*.]

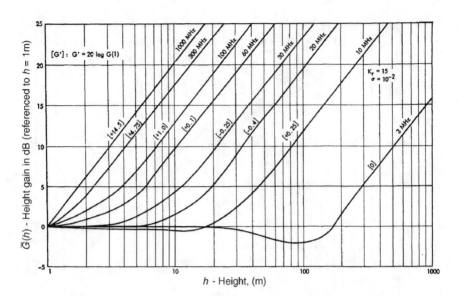

Figure 3.10 Height gain for vertical polarization over good earth. [©*1969 IEEE.*]

height h_1 and the receiving antenna height h_2. That is, the effect of raising the source or the receiving antenna is independent of the other.

At angles of incidence near grazing (small values of angle γ), both ρ_h and ρ_v approach -1 (Fig. 3.5). Also, for small values of angle γ (less than approximately $15°$), the path length difference is well approximated by

$$R_2 - R_1 = 2h_1h_2/r$$

and the phase difference by

$$\beta(R_2 - R_1) = 4\pi h_1 h_2/\lambda r.$$

Then the electric field strength over plane earth, using (3.5), (3.10), and (3.11), reduces to the following simple form:

$$E(r) = 2E_o \sin(2\pi h_1 h_2/\lambda r) \tag{3.14}$$

where $E_o = \sqrt{30P_oD}/r$ is the free-space field strength, from (3.5)
        P_o power radiated by the transmitting antenna
        D directive gain of the transmitting antenna.

When $2\pi h_1 h_2/\lambda r$ is sufficiently small so that

$$\sin(2\pi h_1 h_2/\lambda r) \approx 2\pi h_1 h_2/\lambda r, \tag{3.15}$$

(3.14) further simplifies to

$$E(r) = 4E_o\pi h_1 h_2/\lambda r. \tag{3.16}$$

The attenuation rate of the field in (3.16) is $1/r^2$ (40 dB per decade distance), which is the fall-off rate of the fields in Figs. 3.6 to 3.8 at farther distances.

3.3   PROPAGATION OVER A PERFECTLY CONDUCTING PLANE

A perfectly conducting ground plane is defined as one having infinite conductivity ($\sigma = \infty$). The surface wave does not exist over a perfectly conducting plane. Only the space wave, composed of the direct and reflected waves, is of significance. Over a perfectly conducting plane, $\rho_h = -1$ and $\rho_v = +1$ for all angles of incidence.

The model for propagation over a perfectly conducting plane is applicable to open test ranges with metal ground planes commonly used for antenna calibration and equipment emission measurements. Ground planes are typically constructed from solid sheet metal, metal screens and grids. The size of these ground planes rarely exceeds 100 m in length. The size of the ground screen at the National Institute of Standards and Technology open test site in Boulder, Colorado, for example, is 30 m by 60 m. Another application is propagation calculations in semi-anechoic chambers. Theoretical site attenuation models [11]–[14] are based on field strength calculations over an ideal test site with a perfectly conducting plane of infinite extent.

Horizontal (Perpendicular) Polarization

The case of horizontal polarization (electric field *perpendicular* to the plane of incidence and parallel to the reflecting surface) is relatively straightforward when the transmitting antenna is a horizontal dipole, since the directive gain pattern $D(\Theta)$ is constant in the plane of incidence. That is, the antenna pattern is circular. The directive gain for a short dipole is $D = 1.5$. For a half-wave dipole, $D = 1.64$.

Refer to the geometry in Fig. 3.4. From (3.5), (3.10), and (3.11) with $\rho_h = -1$, the far-field electric field strength is

$$E(r) = \sqrt{30 P_o D} \left[\frac{\varepsilon^{-j\beta R_1}}{R_1} - \frac{\varepsilon^{-j\beta R_2}}{R_2} \right]. \tag{3.17}$$

If the transmitting antenna is a high-gain antenna such as a horn or dish, the directivity in the direction of both the direct wave and the wave incident on the ground plane would have to be accounted for in (3.17). See the treatment for vertical polarization below.

The magnitude of (3.17) is

$$E(r) = \sqrt{30\,P_o D}\frac{[\,R_2^2 + R_1^2 - 2R_1 R_2 \cos \beta (R_2 - R_1)]^{1/2}}{R_1 R_2} \qquad (3.18)$$

where $R_1 = [r^2 + (h_1 - h_2)^2]^{1/2}$
        $R_2 = [r^2 + (h_1 + h_2)^2]^{1/2}.$

For small values of angle γ,

$$R_1 \approx R_2 \approx r$$

$$R_2 - R_1 = 2h_1 h_2/r$$

$$\beta\,(R_2 - R_1) = 4\pi h_1 h_2/\lambda r$$

and (3.18) reduces to

$$E(r) = \frac{2\sqrt{30\,P_o D}}{r}\,\sin\left(\frac{2\pi h_1 h_2}{\lambda r}\right) \qquad (3.19)$$

which is identical to the field over plane earth given by (3.14). The significance is that, for angles of incidence near grazing (low values of angle γ), horizontally polarized fields over a perfectly conducting ground plane are the same as horizontally *and* vertically polarized fields over earth.

Vertical (Parallel) Polarization

Field strength calculations for vertical polarization are somewhat more complicated than for horizontal polarization since the directivity of the transmitting antenna must be accounted for, even for such simple radiators as short dipoles and half-wave dipoles.

Refer to Fig. 3.11. The direct wave and the wave incident on the ground plane are parallel polarized waves, that is, the electric field vector is parallel to the plane of incidence. We wish to calculate only the vertical component of the electric field at the receiving point, E_v.

In Fig. 3.11, the coordinate system for the transmitting antenna is (R, Θ). The sum of the direct and ground-reflected waves using (3.5) and (3.10) with $\rho_v = +1$ is

$$E(r) = \sqrt{30\,P_o}\left[\frac{\sqrt{D(\Theta_1)}}{R_1}\varepsilon^{-j\beta R_1} + \frac{\sqrt{D(\Theta_2)}}{R_2}\varepsilon^{-j\beta R_2}\right]. \qquad (3.20)$$

The vertical component of the direct wave is $E_o(\Theta_1)\sin\Theta_1$, and the vertical component of the reflected wave is $E_o(\Theta_2)\sin\Theta_2$. Then the vertical component of the total field is

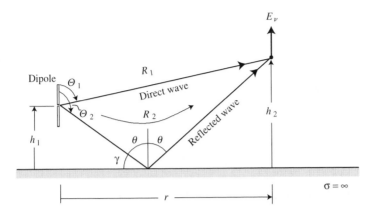

Figure 3.11 Geometry for vertical polarization.

$$E_v(r) = \sqrt{30\,P_o}\left[\frac{\sqrt{D(\Theta_1)}}{R_1}\sin\Theta_1\varepsilon^{-j\beta R_1} + \frac{\sqrt{D(\Theta_2)}}{R_2}\sin\Theta_2\varepsilon^{-j\beta R_2}\right]. \qquad (3.21)$$

Equation (3.21) is the general form for any antenna with directive gain $D(\Theta)$.

For a short dipole antenna, $D(\Theta) = 1.5\sin^2\Theta$ (see, for example, Collin and Zucker [15]), and (3.21) reduces to

$$E_v(r) = \sqrt{45\,P_o}\left[\frac{\sin^2\Theta_1}{R_1}\varepsilon^{-j\beta R_1} + \frac{\sin^2\Theta_2}{R_2}\varepsilon^{-j\beta R_2}\right]. \qquad (3.22)$$

But $\sin\Theta_1 = r/R_1$ and $\sin\Theta_2 = r/R_2$. Then

$$E_v(r) = \sqrt{45\,P_o}\left[\frac{r^2}{R_1^3}\varepsilon^{-j\beta R_1} + \frac{r^2}{R_2^3}\varepsilon^{-j\beta R_2}\right]. \qquad (3.23)$$

The magnitude of (3.23) is

$$E_v(r) = \frac{\sqrt{45\,P_o}\,r^2[R_1^6 + R_2^6 + 2R_1^3 R_2^3\cos\beta(R_2 - R_1)]^{1/2}}{R_1^3 R_2^3} \qquad (3.24)$$

where $R_1 = [r^2 + (h_1 - h_2)^2]^{1/2}$
$R_2 = [r^2 + (h_1 + h_2)^2]^{1/2}.$

For small values of angle γ such that

$$R_1 \approx R_2 \approx r \qquad \text{and} \qquad \cos\beta(R_2 - R_1) \approx 1,$$

(3.24) reduces to

$$E_v(r) = \frac{2\sqrt{45\, P_o}}{r} \tag{3.25}$$

which is twice the free-space field.

Equation (3.25) was calculated for a short dipole transmitting antenna, but it can be generalized for any vertical antenna with directivity D. Thus

$$E_v(r) = \frac{2\sqrt{30\, P_o\, D}}{r}. \tag{3.26}$$

Compare (3.26) with (3.19). As expected, since $\rho_v = +1$, the vertical electric field over a perfectly conducting plane for small values of angle γ is twice the free-space field.

3.4   ATTENUATION OF ELECTROMAGNETIC FIELDS BY BUILDINGS

External electromagnetic fields incident on building structures are attenuated by reflections at the exterior walls and by scattering and reflections inside the building. (Absorption does not play a significant role in building penetration losses, except perhaps, at very high frequencies.) The reduced field-strength levels inside buildings may be desirable or undesirable, depending on the type of signal or noise and the particular application. The attenuation of AM, FM, and TV broadcast signals and of wireless communication signals (e.g., radio paging, cellular, PCS, and cordless phones) results in degraded reception and is undesirable. On the other hand, the attenuation of high-level lightning, microwave and radar fields, for example, would be desirable.

The attenuation of electric and magnetic fields by multistory office buildings, single-story concrete block buildings, and single-family residences was reported in [16]. The results of this study are summarized in Figs. 3.12 to 3.14.

The E field and H field building penetration losses for multistory steel-frame office buildings at a distance of 15 m from the exterior walls is shown in Fig. 3.12. This is a composite of measurements on three buildings:

- a twenty-story office building with marble exterior wall panels
- a four-story office building with preformed concrete walls
- a four-story office building with brick exterior walls.

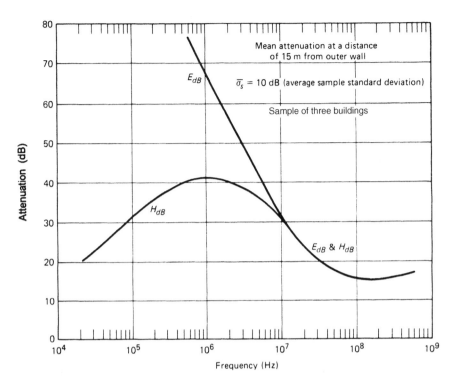

Figure 3.12 Building attenuation for multistory office buildings at a distance of 15 m from the outer wall. [*Copyright ©1985 IEEE. All rights reserved.*]

In Fig. 3.12, $\bar{\sigma}_s$ is the average sample standard deviation for the many measurement sets in these three buildings. The data indicates that the distribution is log normal. As expected, at locations closer to the outer wall, the attenuation is less, while at distances farther than 15 m from the outer wall, the attenuation is greater. This dependence on distance from the outer wall is generally not linear, but depends on the interior design of the building (metal wall panels and partitions, suspended ceilings, duct work, etc.).

The attenuation for commercial single-story concrete block buildings is shown in Fig. 3.13. This type of construction is typical of single-story factory buildings and stores found in shopping centers and malls.

The attenuation for single-family detached residences is shown in Fig. 3.14. Note that below approximately 10 MHz, the magnetic field attenuation is almost zero, while the electric field attenuation reaches a peak of 35 dB around 50 kHz.

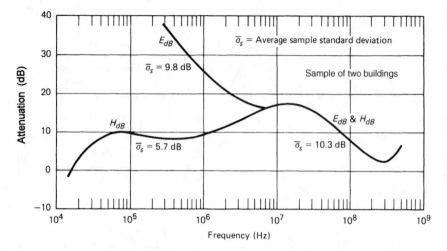

Figure 3.13 Building attenuation for commercial single-story concrete block buildings. [*Copyright © 1985 IEEE. All rights reserved.*]

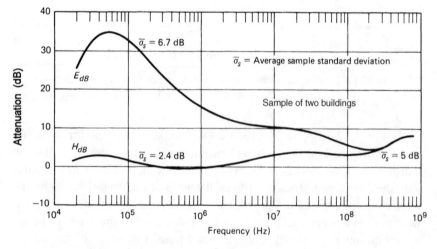

Figure 3.14 Building attenuation for single-family residences. [*©IEEE 1985.*]

Examination of the data in Figs. 3.12 to 3.14 leads to two observations. First, below approximately 10 MHz, the wave impedance inside buildings, defined as the ratio of the electric and magnetic field magnitudes, is less than the intrinsic impedance of free space and decreases with decreasing frequency [16]. This is a consequence of the boundary conditions on the fields. The tangential electric fields are zero on metal boundaries (wiring, pipes, steel partitions, aluminum-foil-backed insulation, metal window frames, appliances, aluminum siding, etc.), while the tangential magnetic fields are twice the incident fields.

The low wave impedance below approximately 10 MHz makes it clear why loop antennas are used in AM radio receivers designed for indoor use.

The second observation is that above approximately 10 MHz, the E field and H field attenuations converge. Evidently, at the higher frequencies, scattering and reflections of both fields are similar.

3.5   EDGE DIFFRACTION

Occasions arise in propagation studies when an estimate of the radio path loss due to an isolated obstacle is required. An example of a natural obstacle is a mountain ridge. In this section, the diffraction losses for a knife edge obstacle and a round obstacle are examined.

Diffraction loss is a critical function of the shape and location of the obstacle and the frequency. The diffraction loss for an actual obstacle may be expected to lie somewhere between the extremes of an ideal knife edge and an ideal round edge.

The diffraction loss in decibels, $A(v^*, \rho^*)$, is plotted in Figs. 3.15 and 3.16 as a function of two dimensionless parameters v^* and ρ^* [17]. The quantities v^* and ρ^* are expressed as functions of

- the obstacle height H above a straight line of length d connecting the transmitter T_x and receiver R_x
- the lengths r_1 and r_2 and the base $d = R_1 + R_2$ of the triangles indicated in Figs. 3.15 and 3.16
- the wavelength λ
- the obstacle radius of curvature r:

$$v^* = H \sqrt{\frac{2}{\lambda} \left[\frac{1}{R_1} + \frac{1}{R_2} \right]}$$

$$\rho^* = 0.261 \lambda^{1/6} r^{1/3} \left[\frac{d}{r_1 r_2} \right]^{1/2} . \tag{3.27}$$

H and thus V^* is negative when the obstacle is below the line connecting the transmitter and receiver.

Insets in Figs. 3.15 and 3.16 aid in the interpretation of the independent variable relationships. The figures apply to both vertical and horizontal polarizations.

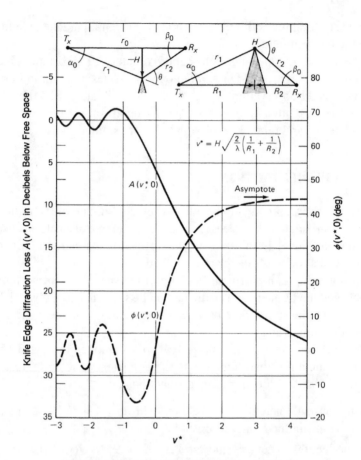

Figure 3.15 Knife-edge diffraction loss, $A(v^*, 0)$.

For the knife edge obstacle in Fig. 3.15, the radius of curvature is $r = 0$, and therefore $\rho^* = 0$. $\phi(v^*, 0)$ in Fig. 3.15 is the phase angle of the diffracted component. Figure 3.15 is identical to the $\rho^* = 0$ curve in Fig. 3.16, except for the change in scale on the abscissa. Note that the diffraction loss for a round edge obstacle with a finite radius of curvature is always greater than the loss for a knife edge.

Figures 3.15 and 3.16 are valid for either linear polarization when the following conditions are satisfied:

- all distances and obstacle dimensions are much larger than a wavelength

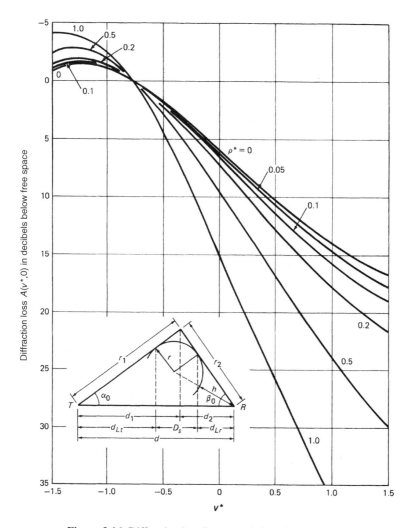

Figure 3.16 Diffraction loss for a round obstacle, $A(v^*, \rho^*)$.

- the dimensions of the diffracting edge perpendicular to the radio path
  equals or exceeds the diameter of the Fresnel zone.

Both conditions are readily satisfied in practice at frequencies above
30 MHz.

For obstacles with a narrow transverse profile (for example, an isolated
building) reflections on transmission paths on each side of the diffracting
edge may result in multipath propagation effects.

Knife-Edge Example

The following numerical example illustrates the use of Fig. 3.15. All dimensions are expressed in meters.

$$f_M = 60 \text{ MHz}$$
$$\lambda = 300/60 = 5 \text{ m}$$
$$H = +915 \text{ m } (3000 \text{ ft})$$
$$R_1 = 32,195 \text{ m } (20 \text{ miles})$$
$$R_2 = 64,390 \text{ m } (40 \text{ miles}).$$

From (3.27), $v^* = 3.95$. The knife-edge diffraction loss from Fig. 3.15 is $A(v^*, 0) = 25$ dB below free space.

3.6   RAYLEIGH ROUGHNESS CRITERION

A wave incident on the surface between a transmitting antenna (source) and a receiving antenna (field point) may be specularly reflected or scattered, depending on the surface roughness, the elevation angle γ, and the wavelength. The path geometry is shown in Fig. 3.17(a), and the surface roughness is depicted as a step discontinuity of height δ in Fig. 3.17(b).

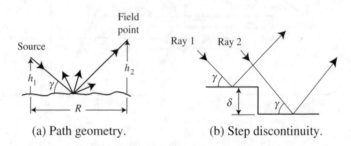

(a) Path geometry.   (b) Step discontinuity.

Figure 3.17   Geometry for surface roughness analysis.

The difference in the path lengths of the two rays in Fig. 3.17(b) is

$$\Delta d = 2\delta \sin \gamma.$$

The phase difference of the two rays is

$$\Delta\phi = \beta \Delta d = \frac{2\pi}{\lambda} \Delta d = \frac{4\pi\delta}{\lambda} \sin \gamma.$$

If the phase difference $\Delta\phi$ is zero, specular reflection occurs and the surface is considered smooth. If the phase difference $\Delta\phi$ is equal to π, the two rays cancel and no energy is transferred in the γ direction. That is, all the energy must be scattered in other directions and the surface is considered rough. The dividing line between smooth and rough was chosen by Rayleigh as $\Delta\phi = \pi/2$.

Using the Rayleigh criterion, a surface is "smooth" if

$$\delta < \frac{\lambda}{8\sin\gamma}.$$

For distance R and heights h_1 and h_2 in Fig. 3.17(a), we have

$$\sin\gamma = \frac{h_1 + h_2}{\sqrt{R^2 + (h_1 + h_2)^2}}.$$

The Rayleigh roughness criterion is then

$$\delta < \frac{\lambda}{8}\frac{\sqrt{R^2 + (h_1 + h_2)^2}}{(h_1 + h_2)}. \tag{3.28}$$

The Rayleigh roughness criterion in (3.28) specifies the maximum allowable surface deviation (from perfectly flat) in order for a surface to still be considered "smooth." This criterion has applications in the design and construction of electromagnetic measuring sites and antenna calibration ranges. See, for example, ANSI C63.7 [18] and Section 4.5 in Chapter 4.

REFERENCES

[1] IEEE Std 100-1977, *IEEE Standard Dictionary of Electrical and Electronics Terms*, published by IEEE, New York, distributed in cooperation with Wiley-Interscience, New York, 1977.

[2] C. R. Burrows, "Radio Propagation Over Plane Earth—Field Strength Curves," *Bell System Technical Journal*, pp. 45–75, January 1937.

[3] C. R. Burrows, "Addendum to 'Radio Propagation Over Plane Earth—Field Strength Curves'," *Bell System Technical Journal*, pp. 574–577, October 1937.

[4] C. R. Burrows and M. C. Gray, "The Effect of the Earth's Curvature on Ground-Wave Propagation," *Proceedings of the IRE*, vol. 29, pp. 16–24, January 1941.

[5] C. R. Burrows, "Addendum to 'The Effect of the Earth's Curvature on Ground-Wave Propagation'," *IEEE Transactions on Antennas and Propagation (Communications)*, vol. AP-12, pp. 789–791, November 1964.

[6] K. A. Norton, "The Calculation of Ground-Wave Field Intensity Over a Finitely Conducting Spherical Earth," *Proceeding of the IRE*, vol. 29, pp. 623–639, December 1941.

[7] K. Bullington, "Radio propagation at frequencies above 30 megacycles," *Proceedings of the IRE*, pp. 1122–1136, October 1947.

[8] CCIR Recommendation 527, "Electrical Characteristics of the Surface of the Earth," *Recommendations and Reports of the CCIR, 1978, Geneva, International Telecommunications Union, vol. V, Propagation in Non-Ionized Media, XIV Plenary Assembly*, Kyoto, 1978.

[9] E. N. Skomal and A. A. Smith, Jr., *Measuring the Radio Frequency Environment*, Van Nostrand Reinhold, Co., New York, 1985.

[10] A. A. Smith, Jr., "Electric Field Propagation in the Proximal Region," *IEEE Transactions on Electromagnetic Compatibility*, vol. EMC-11, no. 4, pp. 151–163, November 1969.

[11] A. A. Smith, Jr., R. F. German, and J. B. Pate, "Calculation of Site Attenuation From Antenna Factors," *IEEE Transactions on Electromagnetic Compatibility*, vol. EMC-24, no. 3, pp. 301–316, August 1982.

[12] A. A. Smith, Jr., "Standard-Site Method for Determining Antenna Factors," *IEEE Transactions on Electromagnetic Compatibility*, vol. EMC-24, no. 3, August 1983.

[13] ANSI C63.4-1991, *Methods of Measurement of Radio-Noise Emissions from Low-Voltage Electrical and Electronic Equipment in the Range of 9 kHz to 40 GHz*, American National Standards Institute, published by the IEEE, SH13896, March 21, 1991.

[14] ANSI C63.5-1988, *American National Standard for electromagnetic compatibility—radiated emission measurements in electromagnetic interference control—calibration of antennas*, American National Standards Institute, published by the IEEE, SH12401, January 23, 1989.

[15] R. E. Collin and F. J. Zucker, *Antenna Theory, part 1*, Inter-University Electronics Series, Vol. 7, McGraw-Hill Book Company, New York, 1969.

[16] A. A. Smith, Jr., "Attenuation of Electric and Magnetic Fields by Buildings," *IEEE Transactions on Electromagnetic Compatibility*, vol. EMC-20, no. 3, August 1978.

[17] P. L. Rice, A. G. Longley, K. A. Norton, and A. P. Barsis, "Transmission Loss Predictions for Tropospheric Circuits," National Bureau of Standards, Tech. Note 101 (revised) January 1967.

[18] ANSI C63.7-1988, *American National Standard—guide for construction of open area test sites for performing radiated emission measurements*, American National Standards Institute, published by the IEEE.

CHAPTER 4

ANTENNAS

This chapter is devoted to a review of the important characteristics of transmitting and receiving antennas. The material in this chapter, much of which was previously presented in [1], is intended for practicing engineers and contains useful information not found in typical texts on antenna theory. The definition of common antenna parameters and terms is presented first, followed by a summary of the interrelationships between many of the antenna parameters, and a discussion of reciprocity. The types of antennas commonly used for measuring electromagnetic fields are reviewed. Last, methods for calibrating receiving antennas are discussed.

4.1 ANTENNA PARAMETERS

The equivalent circuits for receiving and transmitting antennas are shown in Figs. 4.1 and 4.2, respectively. (Also see Figs. 4.3 and 4.4.) In these figures, the impedance, resistance, and reactance of the antennas are denoted as follows:

$Z_A = R_A + jX_A$ antenna impedance
$R_A = R_r + R_L$ antenna resistance
R_r radiation resistance
R_L loss resistance
X_A antenna reactance.

For the receiving antenna shown in Fig. 4.1,

$V_{oc} = h_e E$ open-circuit voltage, V
h_e effective length, meters

69

E	incident electric field strength, V/m
V_R	received voltage (voltage across the load), V
$Z_R = R_R + jX_R$	load impedance (receiver input impedance)
R_R	load resistance
X_R	load reactance.

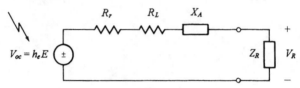

Figure 4.1 Equivalent circuit of a receiving antenna.

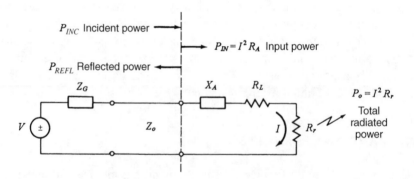

Figure 4.2 Equivalent circuit of a transmitting antenna.

For the transmitting antenna in Fig. 4.2,

V	open-circuit generator voltage, volts
Z_G	generator impedance
P_{INC}	power incident on the antenna input terminals, W
P_{REFL}	power reflected at the antenna terminals, W
$P_{IN} = P_{INC} - P_{REFL}$ antenna input power, W
$P_o = I^2 R_r$ total radiated power, W
I	antenna input current, A
Z_o	characteristic impedance of the transmission line.

Effective Length

The effective length h_e of a receiving antenna is defined as the ratio of the open-circuit terminal voltage to the incident electric field strength in the direction of the antenna's polarization. See Fig. 4.1. It is sometimes referred to as the effective height. The defining relation is

$$h_e = \frac{V_{oc}}{E} \quad \text{m} \tag{4.1}$$

where V_{oc} open-circuit voltage, volts
       E electric field strength, V/m.

Antenna Factor

The antenna factor AF is the ratio of the electric field strength E or magnetic field strength H in the direction of the antenna's polarization to the received voltage V_R. That is (see Fig. 4.1),

$$AF^{\text{electric}} = \frac{E}{V_R} \quad \text{m}^{-1} \tag{4.2}$$

and

$$AF^{\text{magnetic}} = \frac{H}{V_R} \quad \text{siemens/m.} \tag{4.3}$$

The antenna factor includes balun losses, the effective length, and impedance mismatches, but usually does not include transmission line losses.

When applied to the voltage reading of the measuring instrument, the antenna factor yields the electric field strength or magnetic field strength. For example, for electric antennas

$$E = AF V_R \tag{4.4}$$

or, in dB,

$$E(\text{dB}\mu\text{V/m}) = AF(\text{dB/m}) + V_R(\text{dB}\mu\text{V}). \qquad (4.5)$$

If cable loss C_A is not included in the antenna factor (the usual case), it must be added separately. Then

$$E(\text{dB}\mu\text{V/m}) = AF(\text{dB/m}) + V_R(\text{dB}\mu\text{V}) + C_A(\text{dB}). \qquad (4.6)$$

Power Density

The reader should also refer to the discussion of power density in Section 3.1 of Chapter 3, *Free-Space Propagation*. Starting from the basic definition, the power density of a radiator is equal to the real part of the time average Poynting vector:

$$S = \text{Re } \mathbf{P} = \text{Re}(\mathbf{E} \times \mathbf{H}^*).$$

It follows that

$$S(\theta, \phi) = \frac{E^2(\theta, \phi)}{120\,\pi} = 120\pi H^2(\theta, \phi) \qquad (4.7)$$

where $S(\theta, \phi)$ power density, W/m^2
 $E(\theta, \phi)$ RMS electric field strength, V/m
 $H(\theta, \phi)$ RMS magnetic field strength, A/m.

Note: If the *peak* values of the E and H fields are used, the quantities following the equal signs in (4.7) must be multiplied by 1/2, since the RMS fields are $1/\sqrt{2}$ times the peak fields. See (2.81) in Chapter 2.

In the far-field region, E and H vary as $1/r$ and thus S varies as $1/r^2$.

Radiation Intensity

The radiation intensity $U(\theta, \phi)$, also called the antenna power pattern [2], is just the power density multiplied by the square of the radial distance r. Thus,

$$U(\theta, \phi) = r^2 S(\theta, \phi) \qquad \text{watts per steradian (W/sr).} \qquad (4.8)$$

Substituting (4.7) in (4.8), we have

$$U(\theta, \phi) = r^2 \frac{E^2(\theta, \phi)}{120\pi} = 120\pi r^2 H^2(\theta, \phi) \qquad \text{W/sr.} \qquad (4.9)$$

Radiation intensity is the power per unit solid angle radiated by an antenna in the direction (θ, ϕ). Since S falls off as $1/r^2$ in the far field, radiation intensity is independent of distance.

Directive Gain

The directive gain $D(\theta, \phi)$ of an antenna is a measure of the power radiated in different directions. It is similar to the power density $S(\theta, \phi)$, except that distance dependence has been normalized out. That is, while power density falls off as $1/r^2$ in the far field, directive gain is independent of distance.

See (4.8) and (4.9). Directive gain $D(\theta, \phi)$ is defined as 4π times the ratio of the radiation intensity in the direction (θ, ϕ) to the total power radiated by the antenna, P_o. That is,

$$D(\theta, \phi) = 4\pi \frac{U(\theta, \phi)}{P_o}. \tag{4.10}$$

The definition in (4.10) is the directive gain with respect to an isotropic source radiating the same total power, since the radiation intensity of an isotropic source is $U = P_o/4\pi$.

Directive gain is a measure of the directional properties of the radiation pattern normalized to the total radiated power (or output power), and is thus independent of the radiation efficiency or mismatch losses of the antenna.

If the fields from an antenna are known analytically, the directive gain can be calculated. An example is the short dipole. Refer to Section 2.9 in Chapter 2 for the fields from a short dipole.

In the definition of directive gain in (4.10), the radiation intensity is given by (4.9) as

$$U(\theta, \phi) = r^2 \frac{E^2(\theta, \phi)}{120\pi}. \tag{4.11}$$

Using $E(\theta, \phi)$ from (2.68), (4.11) becomes

$$U(\theta, \phi) = \frac{(30\, I\, L\, \beta\, \sin\, \theta)^2}{240\pi}. \tag{4.12}$$

The total radiated power from a short dipole (see Balanis [3]) is

$$P_o = 40\,\pi^2 I_o^2 (L/\lambda)^2. \tag{4.13}$$

Substituting (4.12) and (4.13) into (4.10) yields the directive gain of a short dipole:

$$D(\theta, \phi) = 1.5 \sin^2 \theta. \tag{4.14}$$

Directivity

The directivity D of an antenna is the value of the directive gain in the direction of its maximum value and is a measure of an antenna's ability to concentrate the radiated power in a particular direction. Referring to (4.14), the directivity of a short dipole is $D = 1.5$.

Power Gain

Power gain of an antenna $G(\theta, \phi)$ is defined as 4π times the ratio of the radiation intensity in the direction (θ, ϕ) to the input power to the antenna, P_{IN} (see Fig. 4.2):

$$G(\theta, \phi) = 4\pi \frac{U(\theta, \phi)}{P_{IN}}. \tag{4.15}$$

This is identically the power gain with respect to a lossless isotropic radiator with the same input power. See the remarks under directive gain above.

When the direction is not stated, the power gain G is taken to be the power gain in the direction of maximum radiation.

Power gain is a measure of the directional properties of the radiation pattern normalized to the net input power, and includes the antenna efficiency but not impedance-mismatch losses.

Realized Gain

Realized gain $G_o(\theta, \phi)$ is the power gain of an antenna reduced by the impedance mismatch loss at the antenna input terminals. Realized gain is defined as 4π times the ratio of the radiation intensity in the direction (θ, ϕ) to the incident power P_{INC}. (See Fig. 4.2.) Thus,

$$G_o(\theta, \phi) = 4\pi \frac{U(\theta, \phi)}{P_{INC}}. \tag{4.16}$$

Realized gain accounts for both dissipative losses (radiation efficiency) and impedance mismatch losses.

Total Antenna Efficiency

Total antenna efficiency α_t [3] is the ratio of the total power radiated by an antenna to the incident power from the generator. That is, (Fig. 4.2)

$$\alpha_t = \frac{P_o}{P_{INC}}. \tag{4.17}$$

Total antenna efficiency, also called the *effectiveness ratio*, includes dissipative losses (radiation efficiency) and impedance mismatch losses and is similar in form for transmitting and receiving antennas.

Radiation Efficiency

Radiation efficiency α_r is defined as the ratio of the total power radiated by an antenna to the input power (or net power accepted by the antenna) [4]. See Fig. 4.2:

$$\alpha_r = P_o/P_{IN} = R_r/(R_r + R_L). \tag{4.18}$$

Radiation efficiency is a measure of the dissipative or heat losses in an antenna. Impedance mismatch losses at the input port of the antenna are not included in the radiation efficiency.

Radiation Resistance

The radiation resistance R_r is defined as the ratio of the total power radiated by the antenna P_o to the square of the RMS antenna input current I (see Fig. 4.2):

$$R_r = P_o/I^2 \qquad \text{ohms}. \tag{4.19}$$

This is an equivalent resistance that would dissipate a power equal to the total radiated power when the current through it is equal to the antenna input current [5].

Effective Aperture

The effective aperture of a receiving antenna A_e is defined as the ratio of the received power P_R to the power density S of the incident wave

$$A_e = P_R/S \quad \text{m}^2. \tag{4.20}$$

Maximum Effective Aperture

The maximum effective aperture of a receiving antenna A_{em} is the ratio of the received power under lossless, matched-impedance conditions to the power density of the incident wave. Effective aperture and maximum effective aperture are related by the expression

$$A_e = \alpha_t A_{em} \quad \text{m}^2 \tag{4.21}$$

where α_t is the total antenna efficiency.

Antenna Noise Temperature

Antenna noise temperature T_a is defined by

$$T_a = \frac{p_n}{kB} \quad \text{kelvin (K)} \tag{4.22}$$

where p_n available noise power from an equivalent loss-free
      antenna, W
    k Boltzmann's constant, 1.38×10^{-23} joules/K
    B receiver noise bandwidth, Hz.

External Noise Factor

The noise power received by an antenna may be expressed in terms of an external noise factor f_a defined as

$$f_a = \frac{p_n}{kT_oB} \tag{4.23}$$

where T_o is the reference temperature, 288 K. In decibels,

$$F_a = 10\log_{10} f_a \tag{4.24}$$

or

$$F_a = P_n - 10\log_{10} kT_o - 10\log_{10} B \tag{4.25}$$

where $P_n = 10\log_{10} p_n$, the available noise power in dBW.

Antenna Reflection Coefficient

The antenna reflection coefficient ρ_V is defined as the ratio of the voltage reflected from the antenna input terminals V_{REFL} to the voltage incident on the antenna input terminals V_{INC}, that is,

$$\rho_V = \frac{V_{\text{REFL}}}{V_{\text{INC}}}. \tag{4.26}$$

The reflection coefficient is given in terms of the antenna impedance Z_A and the characteristic impedance of the transmission line Z_o by

$$\rho_V = \frac{Z_A - Z_o}{Z_A + Z_o}. \tag{4.27}$$

Antenna VSWR

The antenna voltage standing wave ratio (VSWR) is defined as the ratio of the maximum to minimum voltage of the standing wave on the transmission line connecting the generator and antenna, assuming that the generator is matched to the transmission line and that the line is lossless. This is denoted as

$$\Gamma = \frac{V^{\max}}{V^{\min}}. \tag{4.28}$$

The antenna VSWR is given in terms of the antenna reflection coefficient magnitude as

$$\Gamma = \frac{1 + |\rho_V|}{1 - |\rho_V|}. \tag{4.29}$$

Conversely, solving (4.29) for $|\rho_V|$, we have

$$|\rho_V| = \frac{\Gamma - 1}{\Gamma + 1}. \tag{4.30}$$

4.2   RELATIONSHIPS BETWEEN ANTENNA PARAMETERS

In many applications, antennas are used mainly for the measurement of electromagnetic field strength. These applications include ambient noise surveys, RF propagation studies, path loss measurements, electromagnetic interference (EMI) measurements, measurements of potentially hazardous radiation levels, and studies of the biological effects of electromagnetic fields.

The effective use of antennas in these applications requires a knowledge of the receiving parameters of antennas—particularly, effective length, antenna factor, effective aperture, and external noise factor. Most antenna manufacturers however, specify antenna parameters that are usually associated with transmission—for example, power gain, directivity, beamwidth, impedance, and VSWR.

In this section, the important relationships between antenna parameters are reviewed, with emphasis on those which relate transmission and reception properties.

Power Gain and Directive Gain

The power gain $G(\theta, \phi)$ and directive gain $D(\theta, \phi)$ of an antenna are related by

$$G(\theta, \phi) = \alpha_r D(\theta, \phi) \qquad (4.31)$$

where α_r is the radiation efficiency. See defining equations (4.10), (4.15), and (4.18).

The maximum power gain G is then given in terms of the directivity by

$$G = \alpha_r D. \qquad (4.32)$$

Realized Gain and Directive Gain

Realized gain $G_o(\theta, \phi)$ and directive gain $D(\theta, \phi)$ of an antenna are related by

$$G_o(\theta, \phi) = \alpha_t D(\theta, \phi) \qquad (4.33)$$

where α_t is the total antenna efficiency. See defining equations (4.10), (4.16), and (4.17).

The maximum realized gain G_o is then given in terms of the directivity by

$$G_o = \alpha_t D. \qquad (4.34)$$

Power Gain and Realized Gain

Since $D(\theta, \phi) = G(\theta, \phi)/\alpha_r$ and $D(\theta, \phi) = G_o(\theta, \phi)/\alpha_t$ from (4.31) and (4.33), respectively, it follows that

$$G_o(\theta, \phi) = \frac{\alpha_t}{\alpha_r} G(\theta, \phi). \qquad (4.35)$$

The ratio α_t/α_r is called the *input impedance mismatch factor*. It can be shown that [1]

$$\frac{\alpha_t}{\alpha_r} = (1 - |\rho_V|^2) = \frac{4R_A Z_o}{(R_A + Z_o)^2 + X_A^2}. \tag{4.36}$$

Then (4.35) can be expressed in terms of the antenna reflection coefficient as

$$G_o(\theta, \phi) = (1 - |\rho_V|^2) \, G(\theta, \phi). \tag{4.37}$$

An equivalent expression in terms of the antenna impedance is

$$G_o(\theta, \phi) = \frac{4R_A Z_o}{(R_A + Z_o)^2 + X_A^2} G(\theta, \phi). \tag{4.38}$$

In summary, (4.37) and (4.38) readily show that realized gain G_o and power gain G differ by the input impedance mismatch factor expressed in terms of either the reflection coefficient or the antenna impedance.

Maximum Effective Aperture, Directivity, and Gain

The maximum effective aperture A_{em} of an antenna is related to the directivity D by [6]

$$A_{em} = \frac{\lambda^2 D}{4\pi} \tag{4.39}$$

where λ is the wavelength.

Since directivity is related to power gain by $D = G/\alpha_r$ and related to realized gain by $D = G_o/\alpha_t$, we also have

$$A_{em} = \frac{\lambda^2 G}{4\pi \alpha_r} = \frac{\lambda^2 G_o}{4\pi \alpha_t} \tag{4.40}$$

where α_r is the radiation efficiency

 α_t is the total antenna efficiency (or effectiveness ratio).

Effective Aperture, Directivity, and Gain

Since effective aperture A_e and maximum effective aperture A_{em} are related by $A_e = \alpha_t A_{em}$, we have from (4.39) and (4.40)

$$A_e = \alpha_t \frac{\lambda^2 D}{4\pi} = \frac{\alpha_t}{\alpha_r} \frac{\lambda^2 G}{4\pi} = \frac{\lambda^2 G_o}{4\pi}. \tag{4.41}$$

It can be seen from these equations that effective aperture is closely related to realized gain G_o since both include impedance mismatch and dissipative losses.

Effective aperture is also given in terms of the power gain G by

$$A_e = (1 - |\rho_V|^2)\frac{\lambda^2 G}{4\pi} \quad (4.42)$$

where $\alpha_t/\alpha_r = (1 - |\rho_V|^2)$ is the input impedance mismatch factor from (4.36).

Power Density, Directivity, and Gain

The following relationships between power density $S(\theta, \phi)$ and directive gain $D(\theta, \phi)$, power gain $G(\theta, \phi)$ and realized gain $G_o(\theta, \phi)$ are readily derived from the definitions of these quantities:

$$S(\theta, \phi) = \frac{P_o D(\theta, \phi)}{4\pi r^2} = \frac{P_{IN}G(\theta, \phi)}{4\pi r^2} = \frac{P_{INC}G_o(\theta, \phi)}{4\pi r^2} \quad (4.43)$$

where r is the radial distance from the antenna, and where P_o, P_{IN}, and P_{INC} are the total radiated power, input power and incident power, respectively. See Fig. 4.2.

Equation (4.43) clearly shows the proper associations of antenna powers (radiated, input, and incident) and antenna gain parameters (directive, power, and realized). This should be helpful to the reader in avoiding a common pitfall—associating a particular gain parameter with the wrong power.

Electric Field and Radiated Power

The power density S of a plane wave in free space is related to the electric field strength by

$$S(\theta, \phi) = \frac{E^2(\theta, \phi)}{120\pi} \quad (4.44)$$

where 120π is the wave impedance or intrinsic impedance of free space. Then,

$$E(\theta, \phi) = \sqrt{120\pi\, S(\theta, \phi)}. \quad (4.45)$$

Substituting (4.43) for $S(\theta, \phi)$ in (4.45) yields the following expressions for the electric field in terms of antenna gains and antenna powers:

$$E(\theta, \phi) = \frac{\sqrt{30P_o D(\theta, \phi)}}{r} = \frac{\sqrt{30P_{IN}G(\theta, \phi)}}{r}$$
$$= \frac{\sqrt{30P_{INC}G_o(\theta, \phi)}}{r} \quad (4.46)$$

where r is the radial distance from the antenna. If the power in (4.46) is in watts, the unit of electric field strength is volts per meter. If the power is in picowatts, E is in microvolts per meter.

Effective Length and Directivity

The power P_R in the load of a matched, lossless receiving antenna ($\alpha_t = 1$) is, by definition, equal to the product of the power density S of the incident field and the maximum effective aperture A_{em} of the antenna. See (4.20) and (4.21). That is,

$$P_R = SA_{em}.$$

But the power density of a plane wave in free space is just

$$S = \frac{E^2}{120\pi}$$

where E is the electric field strength and 120π is the intrinsic impedance of free space.

The maximum effective aperture of an antenna is related to the directivity by (4.39):

$$A_{em} = \frac{\lambda^2 D}{4\pi}.$$

Then

$$P_R = \frac{E^2}{120\pi} \frac{\lambda^2 D}{4\pi}.$$

Refer to the equivalent circuit of a receiving antenna in Fig. 4.1. Since the antenna is assumed lossless, $R_L = 0$. Also, since the antenna is matched, the load resistance R_R is equal to the radiation resistance R_r. Then the voltage across the load is

$$V_R = \sqrt{P_R R_R} = \frac{E\lambda}{2\pi} \sqrt{\frac{D R_r}{120}}.$$

The open-circuit voltage of the antenna V_{oc} is twice the received voltage, that is,

$$V_{oc} = \frac{E\lambda}{\pi} \sqrt{\frac{D R_r}{120}}.$$

But the effective length h_e is defined by (4.1) as

$$h_e = \frac{V_{oc}}{E} \qquad \text{m.}$$

Thus,

$$h_e = \frac{\lambda}{\pi} \sqrt{\frac{DR_r}{120}} \qquad \text{m.} \qquad (4.47)$$

This expression relates the maximum effective length to the directivity and assumes that the major lobe of the antenna is aligned with the direction of arrival of the incident field (and also that the polarization of the antenna is the same as that of the incident field). The effective length for other directions of arrival of the incident field is related to the directive gain $D(\theta, \phi)$ by

$$h_e(\theta, \phi) = \frac{\lambda}{\pi} \sqrt{\frac{D(\theta, \phi) R_r}{120}} \qquad \text{m.} \qquad (4.48)$$

Note that the effective length in (4.47) and (4.48) is a function of directivity or directive gain, radiation resistance and wavelength, but is independent of the load impedance.

Antenna Factor and Realized Gain

Refer again to the equivalent circuit of a receiving antenna in Fig. 4.1. The only assumption we make is that the antenna is terminated in a load resistance R_R. (Since receivers, field strength meters, and spectrum analyzers are designed to have a real input impedance, there is no practical reason to carry the reactive term X_R.)

The magnitude of the received voltage (load voltage) is

$$V_R = \frac{(h_e E) R_R}{[(R_R + R_A)^2 + X_A^2]^{1/2}}.$$

The antenna factor from (4.2) is $AF = E / V_R$. Then

$$AF = \frac{[(R_R + R_A)^2 + X_A^2]^{1/2}}{h_e R_R}.$$

Substituting (4.47) for h_e in this expression yields

$$AF = \frac{\pi}{\lambda R_R} \left[\frac{120}{DR_r} \right]^{1/2} [(R_R + R_A)^2 + X_A^2]^{1/2}.$$

Rearranging the above and multiplying the numerator and denominator by 2 results in

$$AF = \frac{2\pi}{\lambda} \left[\frac{120}{DR_R} \right]^{1/2} \left[\frac{(R_R + R_A)^2 + X_A^2}{4R_R R_r} \right]^{1/2}.$$

The quantity in the right-hand set of brackets is the reciprocal of the total antenna efficiency α_t. Then the above equation can be written

$$AF = \frac{68.83}{\lambda\sqrt{R_R\alpha_t D}}.$$

But $\alpha_t D = G_o$ from (4.34) and we have

$$AF = \frac{68.83}{\lambda\sqrt{R_R G_o}} \tag{4.49}$$

where R_R is the load resistance and G_o is the realized gain.

For a 50-ohm load resistance, (4.49) becomes

$$AF = \frac{9.73}{\lambda\sqrt{G_o}} = \frac{f_M}{30.82\sqrt{G_o}} \tag{4.50}$$

where f_M is the frequency in MHz.

Equation (4.50) relates the maximum antenna factor to the maximum realized gain and assumes that the major lobe of the receiving antenna is aligned with the direction of arrival of the incident field (and, as mentioned previously, that the antenna polarization is the same as that of the incident field). The antenna factor for other directions of arrival of the incident field is related to the realized gain function $G_o(\theta, \phi)$ by

$$AF(\theta, \phi) = \frac{9.73}{\lambda\sqrt{G_o(\theta, \phi)}} = \frac{f_M}{30.82\sqrt{G_o(\theta, \phi)}} \tag{4.51}$$

where a 50-ohm load resistance is assumed. For a different value of load resistance (e.g., a resonant dipole terminated in 73 ohms) use (4.49) with the appropriate directional notation.

Antenna Factor and Power Gain

The relation between realized gain G_o and power gain G is given by

$$G_o = (1 - |\rho_V|^2)\, G$$

where ρ_V is the antenna reflection coefficient. The antenna factor in terms of the power gain is found by substituting the above in (4.49). We have

$$AF = \frac{68.83}{\lambda\sqrt{R_R(1 - |\rho_V|^2)\, G}} \tag{4.52}$$

where R_R is the load resistance. If $R_R = 50$ ohms, this reduces to

$$AF = \frac{9.73}{\lambda\sqrt{(1 - |\rho_V|^2)\, G}} = \frac{f_M}{30.82\sqrt{(1 - |\rho_V|^2)\, G}} \tag{4.53}$$

where f_M is the frequency in MHz.

It was shown in (4.30) that the magnitude of the reflection coefficient is related to the antenna voltage-standing-wave ratio Γ by

$$|\rho_V| = \frac{\Gamma - 1}{\Gamma + 1}. \tag{4.54}$$

Then

$$1 - |\rho_V|^2 = 1 - \frac{(\Gamma - 1)^2}{(\Gamma + 1)^2} = \frac{4\Gamma}{(\Gamma + 1)^2} \tag{4.55}$$

and (4.52) can be written

$$AF = \frac{34.41\,(\Gamma + 1)}{\lambda \sqrt{R_R \Gamma\, G}}. \tag{4.56}$$

If the antenna terminating resistance R_R is 50 ohms, this reduces to

$$AF = \frac{4.87(\Gamma + 1)}{\lambda \sqrt{\Gamma\, G}} = \frac{f_M\,(\Gamma + 1)}{61.64\,\sqrt{\Gamma\, G}}. \tag{4.57}$$

Equation (4.57) may be used to calculate the antenna factor when the power gain and VSWR of the antenna are known.

4.3 RECIPROCITY[1]

The radiated electric field strength of a transmitting antenna can be calculated knowing either the transmitting parameters of the antenna (D, G, G_o) or the receiving parameters of the antenna (AF, h_e). Conversely, when an antenna is used for reception, the incident electric field strength can be calculated from either the transmitting or receiving parameters of the antenna. In this section, the property of reciprocity is reviewed.

The equivalent circuit of a receiving antenna is shown in Fig. 4.3. (Also see Fig. 4.1.) V_R is the received voltage. The input impedance of the receiver is 50 ohms. All balun, cable, attenuator, matching network losses, etc. are included in the box in Fig. 4.3. The antenna and matching networks are assumed to be linear, passive, and bilateral.

By definition,

$$V_R = E/AF \tag{4.58}$$

where AF is the antenna factor of the receiving antenna. Note that all the parameters of the receiving antenna system (effective length, antenna

[1] ©1982 IEEE. Adapted from "Calculation of Site Attenuation from Antenna Factors" by A. A. Smith, Jr., R. F. German and J. B. Pate appearing in *IEEE Transactions on Electromagnetic Compatibility*, vol. EMC-24, no. 3, August 1982, pp. 301–316.

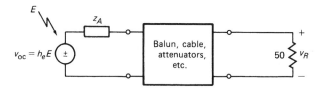

Figure 4.3 Equivalent circuit of a receiving antenna system. [©*1982 IEEE.*]

impedance, and balun, cable, attenuator, and matching network losses) are
embedded in the antenna factor defined in (4.58).

Figure 4.4 shows the equivalent circuit of a transmitting antenna system. (Also see Fig. 4.2.) The 50-ohm source (signal generator or transmitter) has an open circuit voltage V. The antenna current is I. Again, all
balun, cable, attenuator, matching network losses, etc. are relegated to the
box in Fig. 4.4, and the antenna and matching networks are assumed to be
linear, passive, and bilateral.

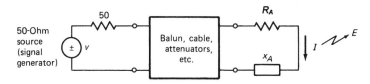

Figure 4.4 Equivalent circuit of a transmitting antenna system. [©*1982 IEEE.*]

For any transmitting antenna, the free-space far-field electric field
strength E at a distance r in the direction of maximum radiation is

$$E = \sqrt{30 P_o D}\ \frac{\varepsilon^{-j\beta r}}{r} \qquad (4.59)$$

which was developed in (4.46).

In (4.59)

P_o total radiated power
D directivity
$\beta = 2\pi/\lambda$ the free-space wave number or phase constant
λ wavelength.

But

$$P_o = I^2 R_r$$

so that

$$E = I \sqrt{30 R_r D}\ \frac{\varepsilon^{-j\beta r}}{r}. \qquad (4.60)$$

The reciprocity theorem will now be used to express the antenna current I of the transmitting antenna in terms of the antenna factor. The antenna factor of the transmitting antenna is the usual antenna factor defined as $AF = E/V_R$ when the antenna is used for reception. Refer to Fig. 4.5 (a), which shows an antenna receiving an electric field E. The received voltage is V_R. By the definition of antenna factor

$$V_R = E/AF. \tag{4.61}$$

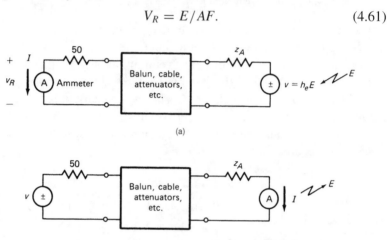

(a)

(b)

Figure 4.5 Illustration of the reciprocity theorem. (a) Antenna receiving an electric field. (b) Same antenna in transmitting mode. [ⓒ*1982 IEEE.*]

Since effective length is defined as $V = h_e E$, (4.61) can be written as

$$V_R = V/h_e AF. \tag{4.62}$$

The current in the 50-ohm termination is

$$I = V_R/50 = V/(50\,h_e AF).$$

But, as shown in (4.47)

$$h_e = \frac{\lambda}{\pi}\sqrt{\frac{DR_r}{120}}.$$

Then the current in the load of the receiving antenna in Fig. 4.5(a) as measured by the ammeter A is

$$I = \frac{V}{50AF}\frac{\pi}{\lambda}\sqrt{\frac{120}{DR_r}}.$$

Invoking the reciprocity theorem, we interchange the ammeter A and voltage source V in Fig. 4.5(a), arriving at Fig. 4.5(b), which shows the antenna in the transmit mode. Evidently, the antenna current for the transmitting antenna in Fig. 4.5(b) is

$$I = \frac{V}{50AF} \frac{\pi}{\lambda} \sqrt{\frac{120}{DR_r}}. \tag{4.63}$$

Substituting (4.63) into (4.60) yields

$$E = \frac{6\pi V}{5\lambda AF} \frac{\varepsilon^{-j\beta r}}{r}.$$

Finally, using $\lambda = 300/f_M$, the free-space far-field electric field strength at a distance r meters can be written as

$$E = \frac{V f_M}{79.58\,AF} \frac{\varepsilon^{-j\beta r}}{r} \tag{4.64}$$

where V open circuit voltage of the 50-ohm signal source
$\quad\quad\; AF$ antenna factor of the transmitting antenna
$\quad\quad\; f_M$ frequency in megahertz.

4.4 TYPES OF RECEIVING ANTENNAS

The types of antennas commonly used for measuring electromagnetic fields in the frequency range of 20 Hz to 40 GHz are summarized in Table 4.1 and described in the remainder of this section. The common attributes of these antennas include small size, lightweight, accurate calibration, and relatively low gain compared to communications antennas.

Small Loop Antennas

Electrically small loops are used to measure electromagnetic fields in the frequency range of approximately 20 Hz to 30 MHz. A loop is considered electrically small if $C_L/\lambda << 1$, where C_L is the length of the perimeter (or circumference for circular loops). The most common loop antenna shapes are circular and rectangular.

While loop antennas respond to the magnetic field component of the wave, the electric field component can also be determined if the wave impedance Z_w is known, since $E = Z_w H$. For plane waves in free-space, $Z_w = 120\pi$.

Loop antennas are usually shielded in order to maintain symmetry and balance. Shields are made of nonmagnetic materials such as aluminum or copper, and have a small air gap.

Table 4.1 Antennas and Probes for Various Frequency Ranges

Antenna Type	Frequency Range
Loop	20 Hz–30 MHz
Vertical monopole	10 kHz–30 MHz
Radiation monitor probes	300 kHz–26 GHz
Broadband dipole (biconical and cylindrical)	30 MHz–200 MHz
Tunable dipole (resonant and halfwave)	30 MHz–1 GHz
Log-periodic (planar and conical spiral)	200 MHz–18 GHz
Discone	1 GHz–10 GHz
Pyramidal horn	1 GHz–40 GHz
Reflector	1 GHz–10 GHz

The effective length h_e of a receiving loop antenna, defined in (4.1), can be derived from Faraday's law. See, for example, (2.3) in Chapter 2. Assuming a sinusoidal time variation and a uniform magnetic field,

$$V_{oc} = -j\omega NAB = -j\omega\mu_o NAH$$

where V_{oc} open-circuit (induced) voltage
$\omega = 2\pi f$
N number of turns
μ_o permeability of free space
A loop area
$B = \mu_o H$.

In the far field, $E = \sqrt{\mu_o/\varepsilon_o}H$. Then

$$|V_{oc}| = 2\pi f NA\sqrt{\mu_o\varepsilon_o} = \frac{2\pi NA}{\lambda} E$$

and

$$h_e = \frac{|V_{oc}|}{E} = \frac{2\pi NA}{\lambda}.$$

The effective length is sufficient to characterize low-frequency loop antennas terminated in a high impedance, that is, when the open-circuit voltage can be measured directly. At higher frequencies, when loop impedance cannot be neglected, or when a matching network or balun transformer is used, the open-circuit voltage cannot be measured directly. In these cases, the antenna factor is required in order to determine the field strength from the received voltage. Since some loop antennas are supplied with magnetic antenna factors while others are supplied with electric antenna factors, care must be taken to be certain that the two are not interchanged when calculating field strengths. See (4.2) and (4.3) for the definitions.

Tunable Dipoles

Tunable dipoles are used for electric field strength measurements from 25 MHz to 1 GHz. There are two types of tunable dipoles—the resonant dipole and the half-wavelength dipole. The resonant dipole is adjusted to a physical length slightly less than one-half wavelength in order to reduce the reactive part of the antenna impedance to zero. The antenna impedance of the halfwave dipole has both real and reactive parts. Rulers are supplied with each of these antennas to adjust the length of the dipole arms for each discrete frequency.

Since the length of a tunable dipole must be adjusted at each frequency, this antenna is only practical for applications involving measurements at a small number of discrete frequencies. In addition, the large physical length of tunable dipoles at the lower frequencies may preclude their use in some applications—for instance, for measurements of vertically polarized fields and in indoor sites, including shielded rooms and anechoic chambers.

The impedance of tunable dipoles is affected by mutual coupling to the ground plane and to nearby metal objects when the distance is less than a wavelength. The impedance of vertical dipoles is also affected by coupling to its transmission line. As a result, the actual antenna factor will differ from the isolated (or free-space) antenna factor. While it is possible to calibrate tunable dipoles to account for mutual coupling under some simple boundary conditions (e.g., fixed height over a large ground plane), it is generally easier to use calibrated broadband dipoles such as biconicals, which exhibit negligible mutual coupling effects because of their small electrical length at the lower frequencies.

One of the historical reasons for the popularity of the thin resonant dipole is the simplicity of the expressions for the effective length and impedance:

$$h_e = \lambda/\pi$$

$$Z_A = 73 + j0.$$

The antenna factor can then be readily calculated. For example, assume the resonant dipole is terminated in 50 ohms. From Fig. 4.1, the received voltage is

$$V_R = \frac{\lambda}{\pi} E \frac{50}{50 + 73}.$$

Using $\lambda = 300/f_M$,

$$AF = \frac{\pi f_M}{300}\frac{123}{50} = 0.02576 f_M$$

or, in dB

$$AF(dB) = 20 \log f_M - 31.8. \tag{4.65}$$

The thin half-wavelength tunable dipole has approximately the same effective length as the resonant dipole, but the antenna impedance is $73 + j42.5$. See King [7] for more exact values of effective length and impedance when the dipole thickness is not negligible.

This ideal analysis neglects the effects of balun transformers, which can cause the actual antenna factor to differ by 2 dB or more from the theoretical value calculated in (4.65). For this reason, the antenna factors of tunable dipoles should be measured if accurate calibration is required. The exception is the "Robert's dipole," which is accurately calibrated by virtue of its design and construction. The Robert's dipole was first described in [8] and has been adopted as the "Reference Antenna" in ANSI Standard C63.5 [9], which gives complete details on the construction and electrical properties of this antenna. Half-wavelength antennas based on Robert's design are available from a number of antenna manufacturers.

Broadband Dipoles

Commercially available biconical antennas and cylindrical dipoles cover the frequency range from 30 to 200 MHz. These antennas are compatible with automated instrumentation and permit swept frequency measurements over their operating range. This can result in a significant reduction in measurement time compared with tunable dipoles which require mechanical adjustment at each frequency. In addition, broadband dipoles, because they are shorter than tunable dipoles at the lower frequencies, are better suited for indoor measurements and for vertical polarization measurements. Because of their smaller size, broadband dipoles exhibit less mutual coupling with nearby metal boundaries compared with tunable dipoles. (A good rule of thumb for broadband dipoles is that mutual coupling will be negligible at 30 MHz if they are placed no closer than 1.5 m from a metal boundary.)

Log-Periodic Antennas

Broadband log-periodic antennas are used to measure electric field strength in the frequency range of 200 MHz–18 GHz. These antennas are moderately directional, with gains in the order of 6–8 dB, and are most commonly used when the direction of the radiation source is known. When the direction of the source is unknown, log-periodic antennas must be scanned in azimuth and elevation to locate the source and align the peak of the major lobe with the incoming wave. In addition, planar log periodic antennas must be rotated on their boresight axes in order to match their polarization to that of the incident wave. If a number of spatially discrete sources must be measured, this could be a time-consuming procedure. However, since the beamwidths of these relatively low directivity antennas are rather broad, exact alignment is not critical and angular scan rates can be high.

Planar log-periodic antennas are used for investigations where it is necessary to measure both orthogonal components (e.g., horizontal and vertical) of elliptically or linearly polarized waves.

Conical log-periodic antennas are used for the measurement of circularly polarized fields of the same sense (right-handed or left-handed), and for measurements of elliptically or linearly polarized fields when it is not necessary to determine the direction of the axis of polarization.

Manufacturers do not typically supply antenna factors for log-periodic antennas. They do, however, specify nominal values of power gain and VSWR and will often provide measured values on request. The antenna factor can then be calculated from

$$AF = \frac{f_M \, (\Gamma + 1)}{61.64 \sqrt{\Gamma \, G}} \qquad (4.66)$$

which was derived in (4.57) for an antenna terminated in 50 ohms. If the antenna is terminated in an impedance other that 50 ohms, (4.56) can be used to calculate the antenna factor.

The following numerical example illustrates the computations required. A log-periodic antenna has a power gain of 8 dB and a VSWR of 2.25:1 at a frequency of 250 MHz. The antenna will be used with a 50-ohm spectrum analyzer. The first step is to convert power gain from decibels to a decimal. Thus,

$$G = 10^{8/10} = 6.31.$$

The antenna factor from (4.66) is

$$AF = \frac{250(2.25 + \ 1)}{61.64 \sqrt{(2.25)(6.31)}} = 3.5.$$

The antenna factor in decibels is

$$AF\,(dB) = 20 \log AF = 20 \log 3.5 = 10.9 \qquad \text{dB.}$$

Vertical Monopoles

Electrically short vertical monopole (or rod) antennas are commonly used to measure vertically polarized electric fields in the range of 10 kHz–30 MHz.

Commercially available monopole antennas come in two standard sizes: 41 in. and 9 ft. The 41-in. rod is preferred for most applications. When measurements of low-level fields are required, the 9-ft rod must be used to attain the necessary sensitivity.

The effective length h_e of a short rod antenna of physical height L is given by $h_e = L/2$. The radiation resistance R_r and the antenna reactance X_A are given by King [7] as

$$R_r = 20\,(\beta L)^2 \quad \text{and} \quad X_A = 198/\beta L$$

where $\beta = 2\pi/\lambda$ is the phase constant.

Commercial rod antennas are coupled to the receiver using either passive matching networks or active broadband pre-amplifiers, the latter having the advantage of a constant antenna factor over the operating frequency range.

Rod antennas must be mounted on a metal ground plane (or imaging plane) for calibrated operation. Failure to do so will change the antenna pattern and antenna impedance (and hence, the antenna factor) in an unpredictable way.

Pyramidal Horns

Octave bandwidth pyramidal horn antennas are used to measure electromagnetic field strengths in the frequency range 1 GHz to 40 GHz. The pyramidal horn is a highly directional antenna with E-plane beamwidths ranging from approximately 40° at the lower frequencies to approximately 9° at the higher end of the frequency range. The beamwidth of a particular antenna depends on design parameters such as flare angle and length. See Balanis [3] for a detailed discussion of the design of horn antennas.

Since they are highly directional, horn antennas are usually used for reception when the location of the radiation source is known. When the location of the source is unknown, horn antennas must be scanned in azimuth and elevation in order to align the major lobe with the direction of the incoming wave. In addition, the polarization of the antenna must be varied to match that of the incident wave. The narrow beamwidths of horn antennas require that angular scanning be done at a rate that is slow enough to ensure that the receiver has time to respond. This could be time consuming, especially if a number of spatially discrete sources are involved.

Horn antennas, by virtue of their narrow beamwidths, do have the advantage of much higher gains compared to omnidirectional antennas. These higher gains might be required, for instance, when low-level fields have to be detected or when high front-end receiver noise figures have to be overcome. Gains of pyramidal horns vary from approximately 10–30 dB over the 1–40-GHz range.

The antenna factor for a horn antenna can be calculated from (4.57) if the power gain and VSWR are known. The following example will illustrate the method. A horn antenna has a power gain of 11.5 dB and a VSWR of 1.14 : 1 at 1000 MHz. The antenna will be used with a 50-ohm spectrum analyzer. The approach here will differ from the example given for log-periodic antennas by immediately taking 20 log of both sides of (4.57). Thus,

$$20 \log AF = 20 \log f_M + 20 \log(\Gamma + 1) - 35.8 - 10 \log \Gamma - G_{dB}$$

or

$$AF(dB) = 20 \log 1000 + 20 \log 2.4 - 35.8 - 10 \log 1.4 - 11.5$$

and

$$AF(dB) = 18.8 \qquad dB.$$

Reflector Antennas

Parabolic reflector (or dish) antennas with horn or log-periodic feeds are used in the 1–40-GHz frequency range when gains higher than those available from horns alone are required. Typical beamwidths of reflector antennas range from 20° at the lower frequencies to 1° at the higher end of the frequency range.

Corresponding power gains range from approximately 15 dB–40 dB from the low- to the high end of the frequency band. See Balanis [3] for a thorough treatment of reflector antennas, especially of reflector directivity and aperture efficiency.

Discone Antennas

Discone antennas are used in the 1- to 10-GHz frequency range for electric field strength measurements. Discones are vertically polarized and omnidirectional in the horizontal plane, having essentially the figure eight pattern of revolution of a vertical dipole. The discone however operates over a much wider frequency range than a tuned dipole.

Radiation Monitor Probes

Radiation monitors are used to measure levels of potentially hazardous electromagnetic fields for comparison to safety standards such as IEEE Standard C95.1-1991 [10] and the recommendations of the Canadian Radiation Bureau of Health and Welfare [11].

The reader is referred to Section 2.13, *Power Density and Hazardous Radiation*, where the measurement of average power density S_{AV} and mean squared electric and magnetic fields E^2 and H^2 is discussed in some detail. The IEEE recommended safety levels of maximum permissible exposure (MPE) are also given in Section 2.13.

Radiation monitor probes measure either the electric component or the magnetic component of the field. Electric field probes use an array of either two or three orthogonal dipoles with diode or thermocouple detectors. (The probes with three orthogonal dipoles have an isotropic radiation pattern.) The frequency ranges covered by electric field probes vary from model to model, but some of the common ranges are 300 kHz–300 MHz, 2–500 MHz, 1 MHz–3 GHz, 300 MHz–18 GHz, 10 MHz–26 GHz, and the ISM (Industrial, Scientific, and Medical) frequencies of 915 MHz and 2450 MHz.

Magnetic field probes use an isotropic array of three orthogonal loop antennas with diode or thermocouple detectors, and cover the frequency range of approximately 3–300 MHz, the exact range depending on the particular model.

Radiation monitors are commonly calibrated in units of equivalent plane-wave power density (in mW/cm^2), though some read field strength directly.

4.5   ANTENNA CALIBRATION

Accurate quantitative measurements of electromagnetic fields require accurate antenna factors. Except for a few simple antenna types, accurately calibrated antenna factors must be established by measurement. (The antenna factors of elementary antennas such as small loops, short dipoles, resonant dipoles, and pyramidal horns can be calculated quite accurately if ideal terminal conditions can be assumed. However, the use of balun transformers or matching networks with these otherwise simple antennas may make antenna factor calculations difficult and unreliable.)

In the remainder of this chapter, four methods for measuring antenna factors are described:

- the standard-antenna method (SAM)
- the standard-field method (SFM)
- the secondary standard-antenna method (SSAM),
- the standard-site method (SSM).

The standard-antenna method (SAM) and standard-field method (SFM) are suitable for use only by standards laboratories such as the National Institute of Standards and Technology (NIST)[2] in the U.S. and national standards laboratories in other countries. Few industrial laboratories have the specialized instrumentation and facilities necessary to use these methods to calibrate antennas. NIST in Boulder, Colorado, provides antenna calibration services for industry for a fee.

Once an antenna has been calibrated (for example, by the standard-antenna method, standard-field method, or standard-site method), it can be used as a secondary standard to calibrate other antennas using the same substitution procedure employed in the standard-antenna method. This secondary standard-antenna method (SSAM) results in some increase in calibration uncertainty but is inexpensive and easy to implement.

The standard-site method (SSM) is the method of choice for many industrial laboratories since it is easy to implement and requires no special instrumentation other than a spectrum analyzer and tracking generator (or field strength meter and signal generator). The SSM requires site attenuation measurements made on an open-field site. Accuracies comparable to those obtained by NIST using the SAM and SFM methods can be achieved. The details for the SSM can be found in American National Standard C63.5 [9] and are summarized later in this section. The theoretical background was developed in Ref. [12].

Standard-Antenna Method

The standard-antenna method (SAM) is used by the National Institute of Standards and Technology (NIST) to calibrate dipole-type antennas (including broadband biconicals, cylindrical dipoles, and log-planar arrays) in the frequency range of 30–1000 MHz and is described in detail by Taggart and Workman [13].

The standard-antenna method is performed at NIST on an open-field site in order to eliminate field nonuniformities due to reflections and scattering from nearby metal objects. A standard receiving antenna consisting of a self-resonant dipole with a balanced detector built into the center is

---

[2] Formerly the National Bureau of Standards (NBS).

placed at a fixed height above ground in the far field of a transmitting dipole. The transmitter power is set to a convenient level, well above ambient noise, and held constant during the measurements. A high impedance voltmeter is used to measure the DC output of the balanced detector, from which the open-circuit induced voltage V_{oc} of the standard dipole is calculated. The electric field strength E which illuminates the standard antenna is calculated from [see (4.1)]

$$E = \frac{V_{oc}}{h_e}$$

where h_e is the effective length of the standard antenna. (The effective length is λ/π times a correction factor which is a function of the length-to-diameter ratio and accounts for the nonsinusoidal current distribution on the dipole.)

The antenna being calibrated is substituted for the standard antenna and the output voltage V_R is measured with a calibrated 50-ohm field strength meter. The antenna factor, AF, of the antenna being calibrated is then calculated as [see (4.2)]

$$AF = \frac{E}{V_R}.$$

NIST places the transmitting and receiving antennas 100 ft apart in the frequency range 30–400 MHz and 30–40 ft apart in the frequency range 400–1000 MHz. Both transmitting and receiving antennas are placed at a fixed height of 10 ft. Horizontal polarization is normally used, but antennas can also be calibrated using vertically polarized fields.

The accuracy of the standard-antenna method depends on the calibration accuracy of the balanced detector built into the center of the standard antenna. This balanced detector consists of a selected point-contact silicon crystal diode and an RC filter network. The detector diode response must be corrected for temperature and for frequency.

The accuracy of the standard-antenna method also depends on how precisely the geometry is duplicated when the antenna being calibrated is substituted for the standard antenna. The error is a function of the spatial rate of change of the field at the position of the antennas. If the antennas are positioned near a null caused by the destructive interference of the direct and ground-reflected waves (Fig. 3.11), the field gradient will be large and small differences in antenna positions can result in large measurement errors. The calibration uncertainty of antenna factors calibrated by NIST using the standard-antenna method is estimated to be within 1 dB or 12 percent.

The standard-antenna method is by necessity a discrete frequency calibration method and is quite time consuming. By way of contrast, the standard-site method (SSM) allows swept frequency measurements which are faster and results in antenna factors which are continuous functions of frequency (i.e., no gaps in the antenna factor spectrum).

Standard-Field Method

The standard-field method (SFM) is used by the National Institute of Standards and Technology to calibrate loop antennas in the frequency range of 30 Hz–50 MHz, monopole antennas in the frequency range of 10 kHz–300 MHz, and horn and dipole-type antennas in the range of approximately 200 MHz–18 GHz.

The standard field is calculated in terms of the known properties of the transmitting antenna, the measured input to the transmitting antenna (current, voltage, or power), and the distance from the transmitting antenna. The antenna being calibrated is placed in the standard field (E or H) and the output voltage of the antenna V_R is measured with an accurate 50-ohm field-strength meter or receiver. The antenna factor of the antenna being calibrated is then calculated from [see (4.2) and (4.3)]

$$AF^{\text{electric}} = \frac{E}{V_R} \qquad \text{m}^{-1}$$

and

$$AF^{\text{magnetic}} = \frac{H}{V_R} \qquad \text{siemens/meter.}$$

The accuracy of the standard-field method depends on a number of factors including the accuracy of the placement of the antenna being calibrated in the spatial position for which the standard field is calculated; the uniformity of the field over the antenna being calibrated; and how well the physical transmitting antenna conforms to the theoretical model from which the field is calculated, the accuracy of the measured input to the transmitting antenna, and the measured output of the receiving antenna.

Loop antennas are calibrated at NIST using a 20-cm diameter standard transmitting loop to generate a standard field. The loop being calibrated and the transmitting loop are placed on a table top, positioned coaxially, and spaced an appropriate distance apart (either 1 or 2 m, depending on the diameter of the receiving loop). The transmitting loop is driven with a current of 100 mA and the output of the receiving loop is measured with a calibrated receiver. The detailed procedure, which is quite involved, is described by Taggart and Workman [13]. NIST estimates that the calibration uncertainty of this method is 0.5 dB or less.

Monopole antennas are calibrated by NIST over a 30 m × 60 m metal ground screen using a transmitting monopole to generate a standard field. In the 15-kHz–30-MHz frequency range, an electrically short monopole made from 2.5 m of 14 AWG copper wire is used as the transmitting antenna. In the 30–300-MHz range, a self-resonant monopole is used. The magnitude of the standard field radiated by the transmitting monopole is calculated in terms of the base current, height, and distance from the transmitting monopole. The uncertainty of monopole calibration at NIST is 1 dB [14].

NIST uses standard-gain horn antennas to generate standard fields in an RF anechoic chamber for calibrating dipole-type antennas in the frequency range of 200–1000 MHz. The field radiated by the standard-gain horn is calculated from (4.46):

$$E = \frac{\sqrt{30 P_{INC} G_o}}{r}$$

where G_o is the known realized gain of the horn antenna, P_{INC} is the measured incident power (or forward power) fed to the horn, and r is the distance to the field point.

The antenna being calibrated is placed precisely at the distance r for which the field has been calculated, and its output voltage V_R is measured with a calibrated field-strength meter. (The distance r must be great enough to ensure that each antenna is in the far field of the other. In the NIST anechoic chamber, r may vary up to 6 m.) The antenna factor is then calculated from

$$AF = \frac{E}{V_R} = \frac{\sqrt{30 P_{INC} G_o}}{r V_R}. \tag{4.67}$$

The calibration uncertainty of antenna factors measured by NIST using the standard-field method with a horn-transmitting antenna in an RF anechoic chamber is 0.75 dB or 9 percent [14].

Transverse electromagnetic cells (TEM cells) [15] and parallel-strip transmission lines provide a convenient way to generate standard fields for calibrating small loop antennas, short dipoles and monopoles, and radiation monitor probes. The electric field E is calculated in terms of the applied voltage V and septum spacing (TEM cell) or plate spacing (parallel-strip line) d_s as

$$E = \frac{V}{d_S}.$$

The magnetic field H is related to the electric field E by the intrinsic impedance of free space, that is,

$$H = \frac{E}{120\pi}$$

or

$$H = \frac{V}{120\pi\, d_S}.$$

The uncertainty of the fields generated by TEM cells is in the order of 1 dB.

Secondary Standard-Antenna Method

The complexity of the standard-antenna and standard-field methods used by the National Institute of Standards and Technology makes it difficult and expensive to duplicate in an industrial laboratory. Antenna factor calibration uncertainties less than 1 dB are achieved by NIST through precise design and construction of the standard antennas (dipoles, loops, monopoles and standard-gain horns), the use of highly accurate instrumentation, and the use of careful measurement methods.

Once an antenna has been calibrated, for example by NIST or by the standard site method (SSM), it can be used as a secondary standard to calibrate other antennas using the same substitution procedure used in the standard-antenna method. This secondary standard-antenna (SSAM) method results in some increase in calibration uncertainty, but is inexpensive and easy to implement.

Measurements must be made on an open-field site. The ground plane may be either earth or metal since the absolute value of the field strength is not required. The geometry is not critical but the transmitting and receiving antennas must be spaced far enough apart so that mutual coupling between them is negligible and far-field conditions obtain. It is preferable to keep the transmitting antenna height low in order to minimize the number of peaks and nulls in the radiation pattern caused by the constructive and destructive interference of the direct and ground-reflected waves. At low frequencies, the receiving antenna height must be great enough to minimize mutual coupling between the antenna and its image in the ground plane. At higher frequencies, the receiving antenna height should be scanned (e.g., from 1 to 4 m) in order to avoid placement in or near a field null, where sensitivity to position is great due to a large field gradient. Suggested geometries are given in Table 4.2.

Table 4.2 Suggested Geometries for Calibrating Antennas Using the Secondary Standard-Antenna Method

Frequency	Geometry		
Range MHz	R (m)	h_1 (m)	h_2 (m)
30–600	30	1.5	4
600–1000	15	0.75	1–4 SCAN

R = distance between transmitting and receiving antennas
h_1 = transmitting antenna height
h_2 = receiving antenna height

The procedure is similar to that described previously for the standard-antenna method. At each frequency, a convenient field strength well above the ambient is obtained by adjusting the output level of the signal source driving the transmitting antenna. A stable signal generator with a maximum output power of 0 dBm will suffice in most ambients for the measurement distances in Table 4.2. Under high ambient noise conditions, the measurement distances can be reduced or a power amplifier can be used. Care must be taken to ensure that the transmitted signal will not interfere with licensed communication services.

The output voltage V_{R_1} of the secondary standard antenna is measured first, using a field strength meter or spectrum analyzer. The antenna being calibrated is then substituted in place of the secondary standard antenna and its output voltage V_{R_2} is measured. It is advisable to swap antennas again and repeat the measurement of V_{R_1} in order to ensure that the signal source output has not drifted. The antenna factor AF_2 of the antenna being calibrated is

$$AF_2(\text{dB/m}) = AF_1(\text{dB/m}) + V_{R_1}(\text{dB}\mu\text{V}) - V_{R_2}(\text{dB}\mu\text{V})$$

where AF_1 (dB/m) is the antenna factor of the secondary standard antenna, in decibels per meter.

As in the standard antenna method, horizontal polarization is normally preferred for calibrating antennas using the secondary standard-antenna method because mutual coupling between the antenna elements and the orthogonal transmission line is negligible. Vertical polarization may be used when calibrating most broadband antennas since transmission line coupling effects are negligible for these antennas due to their short electrical length at the lower frequencies (e.g., biconicals) or low backlobe levels (e.g., log-planars). Vertical polarization should be avoided when calibrating resonant or half-wavelength dipoles at the lower frequencies, unless, of course, it is desired to account for the effects of transmission line coupling on the antenna factor.

With proper choice of instrumentation and measurement technique, it is estimated that the added calibration uncertainty introduced by the secondary standard-antenna method can be held to 1 dB or less. Since the calibration uncertainty of NIST-calibrated antennas used as secondary standards is estimated to be 1 dB, the overall uncertainty of antenna factors calibrated by the secondary standard-antenna method can be 2 dB or less.

Standard-Site Method[3]

The standard-site method (SSM) for calibrating antennas is based on site attenuation measurements made on an open-field site. Measurement details can be found in American National Standard C63.5 [9]. The theoretical background was developed in [12]. Also, see [16] for further details on site attenuation measurements. The SSM method is reviewed in this section.

The standard-site method provides a fast, convenient way of calibrating half-wave dipoles, broadband cylindrical dipoles, biconicals, log-planars, horns, and similar types of linearly polarized antennas in the frequency range of approximately 25–1000 MHz. Vertical monopoles have also been calibrated using the SSM.

The accuracy of antenna factors obtained by the standard-site method can approach those achieved by NIST with the standard-antenna and standard-field methods if the requisite site attenuation measurements are made on a flat open-field site devoid of nearby scatterers such as trees, power lines, fences, and vehicles. In practice, near-ideal sites are not difficult to find. An acceptable site for calibration at 10 m is one comprised of a large paved parking lot with a 7 m × 14 m temporary ground plane constructed of overlapped and taped sections of paper barrier insulating foil. The edge of the ground plane should be 20 m or greater from the nearest obstacle [9]. An example is shown in Fig. 4.6.

The standard-site method has the advantage of requiring neither the availability of a standard antenna nor the generation of a standard field. In addition, unlike the standard-antenna method, absolute voltage or field-strength measurements are not required, and unlike the standard-field method, the absolute level of the signal source need not be known. However, a signal source and receiver which are stable over the measurement interval are required. The method is easy to implement and requires no spe-

[3]©1982 IEEE. Adapted from "Standard-Site Method for Determining Antenna Factors," by A. A. Smith, Jr., appearing in the *IEEE Transactions on Electromagnetic Compatibility*, vol. EMC-24, no. 3, August 1982, pp. 316–322.

Figure 4.6 Antennas shown over a metal foil ground plane.

cial instrumentation other than a spectrum analyzer and tracking generator
(or field-strength meter and signal generator).

Description of the Method.   Figure 4.7 shows a pair of antennas separated by a distance R and located over a ground plane with conductivity σ

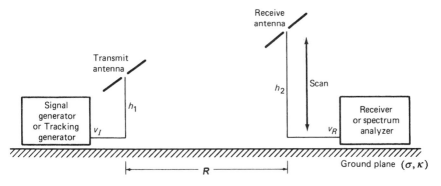

Figure 4.7 Site attenuation measurement setup.

and relative dielectric constant κ. A signal generator or tracking generator with output voltage V_I drives the transmitting antenna, which is placed at a height h_1 above the ground plane. The height h_2 of the receiving antenna is scanned, and the maximum output voltage V_R is measured by a receiver or spectrum analyzer. The purpose of scanning the receiving antenna height is to avoid nulls in the ground-wave field. The *measured* site attenuation is defined as

$$A = V_I / V_R. \tag{4.68}$$

The *theoretical* site attenuation for an *ideal* site (defined as a plane homogeneous surface of infinite extent) is given by

$$A = \frac{279.1\, AF_T\, AF_R}{f_M E_D^{\max}} \tag{4.69}$$

where AF_T antenna factor of transmitting antenna
        AF_R antenna factor of receiving antenna
        f_M frequency, MHz
        $E_D^{\max}$ calculated maximum electric field strength in the receiving antenna height scan range $h_2^{\min} \le h_2 \le h_2^{\max}$ from a half-wave dipole with 1 picowatt of radiated power, in microvolts per meter.

While (4.69) is applicable for either horizontal or vertical polarization, as a practical matter horizontal polarization is most commonly used for antenna calibration measurements. Horizontal polarization is preferred because mutual coupling between the antenna elements and the orthogonal transmission line is negligible, as is scattering from the transmission line. In addition, horizontal polarization is relatively insensitive to site anomalies, including ground-screen edge reflections, and provides results which closely approximate those obtainable in free space.

Electrically small antennas, biconical dipoles, broadband dipoles, and log-periodic arrays designed for use above approximately 100 MHz all have antenna factors which are independent of height and polarization if they are at least 1 m above the ground plane. These antennas can be calibrated in horizontal polarization on a standard site, and the resultant antenna factors will be valid for vertical polarization [9].

Calibration over a metal ground plane ($\sigma = \infty, \kappa = 1$) is preferred for a number of reasons. First, calculations of $E_D^{\max}$ are simplified since the magnitude of the horizontal and vertical reflection coefficients is unity. Also, uncertainties about soil conductivities and permittivities are eliminated.

Values of $E_D^{\max}$ for some typical geometries for horizontal polarization over a metal ground plane are given in Table 4.3. See [9] for additional geometries and for values of $E_D^{\max}$ for vertical polarization over a metal ground plane. See [12] and [16] for values of $E_D^{\max}$ over earth for both horizontal and vertical polarizations.

Measurements. The standard-site method requires three site attenuation measurements made under identical geometries (h_1, h_2, R) using three different antennas taken in pairs as shown in Fig. 4.8. The three equations associated with the three site attenuation measurements are, from (4.69)

$$AF_1 AF_2 = \frac{f_M E_D^{\max}}{279.1} A_1 \tag{4.70}$$

$$AF_1 AF_3 = \frac{f_M E_D^{\max}}{279.1} A_2 \tag{4.71}$$

$$AF_2 AF_3 = \frac{f_M E_D^{\max}}{279.1} A_3 \tag{4.72}$$

where AF_1, AF_2, AF_3 are the antenna factors of antennas 1, 2, and 3
A_1, A_2, A_3 are the measured site attenuations (see Fig. 4.8).

Solving the simultaneous equations (4.70) to (4.72) and expressing the results in decibels gives the desired expressions for the antenna factors in terms of the ground-wave field-strength term and measured site attenuations:

$$AF_1(\text{dB}) = 10 \log f_M - 24.46$$
$$+ \frac{1}{2}[E_D^{\max}(\text{dB}\mu\text{V/m}) + A_1 + A_2 - A_3] \tag{4.73}$$

Table 4.3 Tabulations of E_D^{MAX} for Some Typical Geometries
for Horizontal Polarization over a Metal Ground Plane

R meters	10	10	30	30
h_1 meters	1	2	1	2
h_2 meters	1–4	1–4	1–4	1–4
f_M MHz		E_D^{max} (dBμV/m)		
30	−10.4	−4.8	−28.3	−22.3
35	−9.1	−3.6	−27.0	−21.0
40	−8.0	−2.6	−25.8	−19.9
45	−7.0	−1.7	−24.8	−18.9
50	−6.1	−0.9	−23.9	−18.0
60	−4.7	0.2	−22.3	−16.4
70	−3.5	1.1	−21.0	−15.2
80	−2.4	1.7	−19.9	−14.1
90	−1.6	2.0	−18.8	−13.1
100	−0.8	2.2	−18.0	−12.3
120	0.4	2.4	−16.4	−10.9
125	0.6	2.4	−16.1	−10.6
140	1.2	2.5	−15.1	−9.8
150	1.5	2.5	−14.6	−9.3
160	1.8	2.6	−14.0	−8.9
175	2.1	2.6	−13.3	−8.4
180	2.1	2.6	−13.1	−8.2
200	2.3	2.6	−12.3	−7.7
250	2.5	2.7	−10.6	−6.9
300	2.6	2.7	−9.3	−6.7
400	2.7	2.7	−7.7	−6.7
500	2.8	2.6	−6.8	−6.6
600	2.8	2.6	−6.7	−6.6
700	2.8	2.7	−6.7	−6.6
800	2.8	2.7	−6.7	−6.6
900	2.6	2.7	−6.6	−6.6
1000	2.7	2.7	−6.6	−6.6

$$AF_2(\text{dB}) = 10\log f_M - 24.46$$
$$+ \frac{1}{2}[E_D^{\text{max}}(\text{dB}\mu\text{V/m}) + A_1 + A_3 - A_2] \qquad (4.74)$$

$$AF_3(\text{dB}) = 10\log f_M - 24.46$$
$$+ \frac{1}{2}[E_D^{\text{max}}(\text{dB}\mu\text{V/m}) + A_2 + A_3 - A_1]. \qquad (4.75)$$

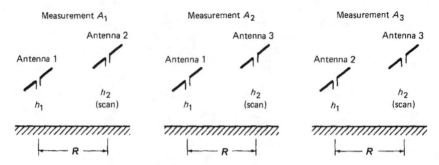

Figure 4.8 Three site attenuation measurements using three different antennas in pairs.

If two identical antennas are to be calibrated, their antenna factor $AF(\text{dB})$ can be obtained from a single site-attenuation measurement A using the following expression:

$$AF(\text{dB}) = 10\log f_M - 24.46 + \frac{1}{2}[E_D^{\max}(\text{dB}\mu\text{V/m}) + A]. \quad (4.76)$$

In practice, two antennas are never identical and the antenna factor calculated by (4.76) will be the geometric mean of the two antennas.

The accuracy of antenna factors determined by the standard-site method from (4.73)–(4.75) depends on the quality of the measuring site and on the accuracy of the site attenuation measurements.

Site quality refers to how closely the measuring site conforms to an ideal site; specifically, how closely the actual ground-wave field strength compares with the theoretical ground-wave field strength over a homogeneous surface of infinite extent.

While it is impossible to quantitatively relate field-strength errors to the physical characteristics of the measuring site, guidelines on what constitutes an acceptable site do exist. Reference [17], for instance, specifies the minimum size of the obstruction-free area, maximum surface roughness derived from Rayleigh's criterion (see Section 3.6 in Chapter 3), and the minimum size and location of the reflecting surface determined by the first Fresnel ellipse.

Site attenuation measurement errors in (4.73)–(4.75) can be minimized by judicious selection of the measurement method since site attenuation is just a measure of the ratio V_I/V_R. Recommended methods for discrete and swept frequency measurements are given in [9].

Accurate antenna calibration also requires some restrictions on measurement geometry. The antenna separation distance R must be great enough to ensure that near-field effects and antenna-to-antenna mutual coupling effects are negligible. Antenna heights (h_1, h_2) must be great enough to minimize antenna-to-ground-plane mutual impedances and to ensure negligible contribution from the surface-wave component of the ground wave. (The surface-wave component does not exist over a metal ground plane.) Scanning of the receiving antenna height h_2 is a practical requirement which eliminates the sensitivity of measurements to nulls. (The large spatial rate of change of the field in the region of a null can result in large measurement errors from small errors in antenna positioning.) Fixed receiving antenna heights may be used for geometries and frequencies where nulls are absent.

An Example.   A cross-polarized log-periodic antenna and two planar log-periodic antennas were calibrated by the standard-site method. The requisite site attenuation measurements were made on a 30 m × 55 m metal ground screen at a distance of $R = 10$ m and antenna heights of $h_1 = 2$ m and $h_2 = 1 - 4$ m.

The site attenuation measurements were made using the swept frequency method. Instrumentation consisted of an HP 8581A spectrum analyzer and an HP 8444A tracking generator under control of an HP 9836 computer/controller. Vertical scanning of the motorized mast was also computer controlled.

Figure 4.9 shows the site attenuation measurement for one of the pairings of the three antennas. This measurement and the site attenuation

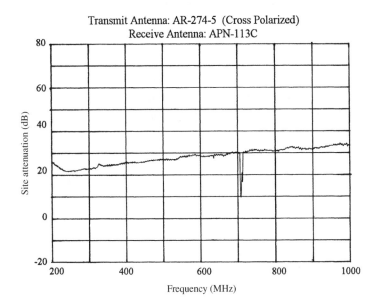

Figure 4.9 Site attenuation measurement for one antenna pair.

measurements for the two other antenna pairings were stored in the computer, and software was used to solve (4.73) to (4.75).

The antenna factor for the cross-polarized log-periodic antenna is shown in Fig. 4.10. Both measured and computer-averaged plots are evident in the figure.

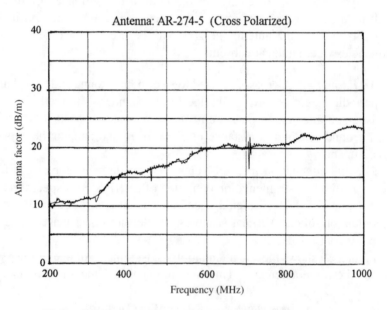

Figure 4.10 Antenna factor plot for the cross-polarized antenna.

The advantage of using swept-frequency rather than discrete-frequency measurements is illustrated in Fig. 4.10. Subtle antenna factor variations readily apparent on the continuous frequency plots could be completely missed with discrete frequency measurements.

REFERENCES

[1] E. N. Skomal and A. A. Smith, Jr., *Measuring the Radio Frequency Environment*, Van Nostrand Reinhold, Co., New York, 1985.

[2] J. D. Kraus and K. R. Carver, *Electromagnetics*, second edition, McGraw-Hill Book Co., New York, 1973.

[3] C. A. Balanis, *Antenna Theory—Analysis and Design*, Harper & Row Publishers, New York, 1982, p. 104.

[4] IEEE Std 100-1977, *IEEE Standard Dictionary of Electrical and Electronics Terms*, second edition, The Institute of Electrical and Electronics Engineers, Inc., New York, distributed in cooperation with Wiley-Interscience.

[5] R. E. Collin and F. J. Zucker, *Antenna Theory—Part 1*, McGraw-Hill Book Co., New York, Toronto and London, 1969.

[6] J. D. Kraus, *Antennas*, McGraw-Hill Book Co., New York, Toronto, and London, 1950.

[7] R. W. P. King, *The Theory of Linear Antennas*, Harvard University Press, Cambridge, MA, 1956.

[8] W. K. Roberts, A New Wide-Band Balun, *Proceedings of the IRE*, December 1957, pp. 1628–1631.

[9] ANSI C63.5-1988, *American National Standard for Calibration of Antennas Used for Radiated Emission Measurements in Electromagnetic Interference (EMI) Control*, The Institute of Electrical and Electronics Engineers, Inc., New York.

[10] IEEE C95.1-1991, *IEEE Standard for Safety Levels with Respect to Human Exposure to Radio Frequency Electromagnetic Fields, 3 kHz to 300 GHz*, Institute of Electrical and Electronics Engineers, Inc., New York, 1992.

[11] *Safety Code 6, Recommended Safety Procedures for the Installation and Use of Radio Frequency and Microwave Devices in the Frequency Range 10 MHz to 300 MHz*, Radiation Protection Bureau of Health and Welfare, Canada, February 1979.

[12] A. A. Smith, Jr., "Standard Site Method for Determining Antenna Factors," *IEEE Transactions on Electromagnetic Compatibility*, vol. EMC-24, no. 3, August 1982, pp. 316–321.

[13] H. E. Taggart and J. L. Workman, "*Calibration Principles and Procedures for Field Strength Meters (30 Hz to 1 GHz)*," U.S. Dept. of Commerce, National Bureau of Standards, Technical Note 370, March 1969.

[14] E. B. Larsen, "*E-Field Antenna Measurements and Calibration—Vertical Monopole Measurements and Calibration*," National Bureau of Standards, lecture notes.

[15] M. L. Crawford, "Generation of Standard EM Fields Using TEM Transmission Cells," *IEEE Transactions on Electromagnetic Compatibility*, vol. EMC-16, no. 4, November 1974, pp. 189–195.

[16] A. A. Smith, Jr., R. F. German, and J. B. Pate, "Calculation of Site Attenuation from Antenna Factors," *IEEE Transactions on Electromagnetic Compatibility*, vol. EMC-24, no. 3, August 1982, pp. 301–316.

[17] ANSI C63.7-1988, *American National Standard—Guide for Construction of Open-Area Test Sites for Performing Radiated Emission Measurements*, The Institute of Electrical and Electronics Engineers, Inc., New York.

CHAPTER **5**

THE RF ENVIRONMENT

Electromagnetic noise and interference limit the performance of all communication systems. The minimum signal levels that can be detected are limited by both external and internal noise sources. The external electromagnetic noise environment is composed of natural and man-made sources. Sources of internal noise in receiving systems include antenna and transmission line losses and noise generated in the receiver.

While noise is completely described by its spatial, spectral, and temporal characteristics, this chapter concentrates mainly on the spectral component—specifically, power spectral density, which is the most basic noise measure.

The first topic in this chapter is a review of common noise parameters and their interrelationships. The receiving system is analyzed next. Natural sources of ambient electromagnetic fields are reviewed, including extraterrestrial (galactic and solar) radio emissions and atmospheric noise. The man-made radio noise spectrum is discussed next, followed by an examination of powerline conducted noise. Last, data on the earth's magnetic and electric fields are presented.

5.1 NOISE PARAMETERS

There are several parameters or measures commonly used to describe electromagnetic noise. The particular choice depends on the origin and characteristics of the noise, the application, and historical precedents.

These parameters fall into two categories: electromagnetic field quantities (power density and field strength) and received quantities (power and voltage). The quantities pertinent to the noise data presented in this chapter are summarized in Table 5.1 and discussed in the remainder of this section.

Table 5.1 SOME COMMON NOISE PARAMETERS

Symbol	Description	Units[*]
s, S	Noise field power spectral density ($S = 10 \log s$)	$W/m^2/Hz$
e_n, E_n	Noise field spectral intensity ($E_n = 20 \log e_n$)	$V/m/Hz^{1/2}$
e, E	Noise field strength ($E = 20 \log e$)	V/m
f_a, F_a	Effective antenna noise factor ($F_a = 10 \log f_a$)	dimensionless
T_a	Effective antenna noise temperature	K
p, P	Noise power spectral density ($P = 10 \log p$)	W/Hz
p_R, P_R	Available noise power ($P_R = 10 \log p_R$)	W
v, V	Noise voltage spectral intensity ($V = 20 \log v$)	$V/Hz^{1/2}$
v_R, V_R	Available noise voltage ($V_R = 20 \log v_R$)	V

[*] Common units of power are watts, milliwatts, and microwatts.
  Common units of voltage are volts, millivolts, and microvolts.
  Bandwidth is commonly expressed as Hz, kHz, and MHz.

NOISE POWER PARAMETERS

If s is the noise field power spectral density incident on a receiving antenna, the noise power spectral density at the output terminals of the antenna is

$$p = sA_e = kT_a \qquad W/Hz \qquad (5.1)$$

where p noise power spectral density
  s noise field power spectral density
  A_e effective aperture of the antenna, m^2. See (4.20).
  k = Boltzmann's constant = 1.38×10^{-23}, Joules/Kelvin
  T_a effective antenna noise temperature in the presence of an external noise field.

The available noise power p_R at the antenna output terminals is

$$p_R = pb = sA_eb = kT_ab \qquad W \qquad (5.2)$$

where b bandwidth in Hz. (Noise power is proportional to bandwidth.)

The effective antenna noise factor f_a is commonly used as a measure of atmospheric and man-made noise power received by an antenna. However, in many applications, f_a is not the most convenient noise measure and must be converted to a more suitable parameter.

f_a is defined as the ratio of noise power available from a loss-free antenna to the noise power generated by a resistor at the reference temperature T_o. That is, from (5.2), using the *maximum* effective aperture A_{em} [see (4.21)] since f_a is defined for a *lossless* antenna, we have

$$f_a = \frac{p_R}{kT_o b} = \frac{p}{kT_o} = \frac{s A_{em}}{kT_o} = \frac{T_a}{T_o} \tag{5.3}$$

where

$$F_a = 10 \log f_a \tag{5.4}$$

and where

p_R available noise power from an equivalent loss-free antenna
$kT_o b$ noise power available in a bandwidth b from a resistor at temperature T_o
T_a effective antenna noise temperature in the presence of an external noise field
T_o $= 288$ K, reference temperature.

Note that f_a is dimensionless, and that both f_a and T_a are independent of bandwidth.

The significance of the definition of f_a in (5.3) is that, given a value of f_a at a particular frequency, the available noise power p_R and noise power spectral density p can be determined. These noise parameters are more convenient to use in many receiving applications. From (5.3), we have

$$p_R = f_a k T_o b \tag{5.5}$$

$$P_R = F_a + 10 \log b - 204 \qquad \text{dBW} \tag{5.6}$$

and

$$p = f_a k T_o \tag{5.7}$$

$$P = F_a - 204 \qquad \text{dBW/Hz} \tag{5.8}$$

where $10 \log k T_o = -204$.

Equations (5.5) and (5.7) reveal why F_a, while dimensionless, is commonly expressed in units of dB above $kT_o b$ or dB above kT_o.

Noise Field-Strength Parameters

P_R and P in (5.6) and (5.8) are available power quantities at the output of a *loss-free* antenna expressed in terms of the effective antenna noise factor F_a. In many applications, it is desirable to express the noise field parameters E_n and E in terms of F_a. Once this is done, the available voltage parameters V and V_R can be determined for *any* receiving antenna in terms of its antenna factor.

By definition,

$$s = \frac{e_n^2}{120\pi} \qquad \text{W/m}^2\text{/Hz.} \tag{5.9}$$

Then

$$e_n = \sqrt{120\pi \, s} \qquad \text{V/m/}\sqrt{\text{Hz.}} \tag{5.10}$$

From (5.3)

$$s = \frac{f_a k T_o}{A_{em}}. \tag{5.11}$$

Substituting (5.11) into (5.10) yields

$$e_n = \sqrt{\frac{120\pi f_a k T_o}{A_{em}}} \qquad \text{V/m/}\sqrt{\text{Hz}} \tag{5.12}$$

and

$$E_n = 20 \log e_n = F_a + 25.76 - 204 - 10 \log A_{em}. \tag{5.13}$$

Published values of F_a are referenced to a short lossless vertical monopole which has a maximum effective aperture of

$$A_{em} = 0.0595\lambda^2 = 0.0595 \left(\frac{300}{f_M}\right)^2 \tag{5.14}$$

where f_M is the frequency in MHz.

Then

$$10 \log A_{em} = 37.2 - 20 \log f_M. \tag{5.15}$$

Substituting (5.15) into (5.13) gives the final result for E_n, the noise field spectral intensity (expressed in units of dB referenced to one microvolt per meter per square root of hertz):

$$\boxed{E_n = F_a + 20 \log f_M - 95.5 \qquad \text{dB}\mu\text{V/m/}\sqrt{\text{Hz.}}} \tag{5.16}$$

Noise field strength e is related to e_n by

$$e = e_n\sqrt{b} \tag{5.17}$$

or

$$E = 20 \log e = E_n + 10 \log b. \tag{5.18}$$

Substituting (5.16) into (5.18) yields the noise field strength (in dB referenced to one microvolt per meter):

$$E = F_a + 20\log f_M + 10\log b - 95.5 \qquad \text{dB}\mu\text{V/m} \qquad (5.19)$$

where b = bandwidth in Hz.

Received Voltage Parameters

When radiated electromagnetic noise is expressed in terms of the noise field parameters E_n and E, the available noise voltage parameters V and V_R are readily calculated by applying the antenna factor of the receiving antenna. Antenna factor is defined in (4.2). Also see equations (4.49) to (4.57).

The available noise voltage spectral intensity at the output of a receiving antenna in terms of the noise field-strength spectral intensity is

$$v = \frac{e_n}{AF} \qquad V/\sqrt{\text{Hz}} \qquad (5.20)$$

or

$$V = 20\log v = E_n - AF(\text{dB/m}) \qquad (5.21)$$

where $AF(\text{dB/m}) = 20\log AF$, antenna factor in dB per meter.

From (5.16), the available noise voltage spectral intensity in terms of the effective antenna noise factor F_a is

$$V = F_a + 20\log f_M$$
$$-AF(\text{dB/m}) - 95.5 \qquad \text{dB}\mu\text{V}/\sqrt{\text{Hz}}. \qquad (5.22)$$

Similarly, the available noise voltage at the output of a receiving antenna in terms of the noise field strength is

$$v_R = \frac{e}{AF} \qquad V \qquad (5.23)$$

or

$$V_R = 20\log v_R = E - AF(\text{dB/m}). \qquad (5.24)$$

From (5.19), the available noise voltage in terms of the effective antenna noise factor F_a is

$$V_R = F_a + 20\log f_M + 10\log b$$
$$-AF(\text{dB/m}) - 95.5 \qquad \text{dB}\mu\text{V}. \qquad (5.25)$$

Noise voltage spectral intensity v and noise voltage v_R are, of course, related by

$$v_R = v\sqrt{b} \tag{5.26}$$

and

$$V_R = V + 10\log b. \tag{5.27}$$

5.2   THE RECEIVING SYSTEM

In the most general case, a receiving system consists of an antenna, an antenna coupling circuit (matching networks, balun transformers), a transmission line, and the receiver, as shown in Figs. 5.1 and 5.3. The performance of the complete receiving system is commonly expressed in two different ways, depending on the application. Both methods account for the external ambient noise as well as noise generated within in the receiving system. The first method characterizes the receiving system performance in terms of its *system noise factor* f_s, which is referenced to the antenna terminals. The second method expresses receiving system performance in terms of the *received noise* referenced to the input of the receiver. Received noise includes both external ambient noise and internal receiver noise. Both methods are discussed below.

System Noise Factor

Refer to Fig. 5.1. Following the development by Spaulding and Disney (see [1]), the noise factor of the antenna circuit is

$$f_c = 1 + \frac{T_c}{T_o}(l_c - 1) \tag{5.28}$$

and the noise factor of the transmission line is

$$f_t = 1 + \frac{T_t}{T_o}(l_t - 1) \tag{5.29}$$

where f_c noise factor of the antenna circuit
  f_t noise factor of the transmission line
  l_c loss factor of the antenna and associated coupling circuit, ≥ 1
  l_t loss factor of the transmission line, ≥ 1
  T_c actual temperature of the antenna circuit and surrounding ground, K
  T_t transmission line temperature, K
  T_o reference temperature, 288 K.

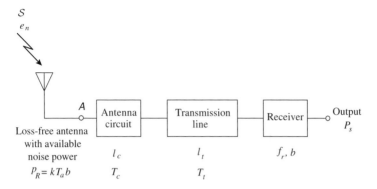

Figure 5.1 The receiving system and available noise power.

The component losses l_c and l_t are defined as the ratio of the power input to the circuit element to the power measured at the output, under matched impedance conditions. Since the antenna coupling circuit and the transmission line are passive elements, both l_c and l_t are always equal to or greater than unity. The component losses in decibels are

$$L_c = 10 \log l_c \qquad \text{dB} \tag{5.30}$$

and

$$L_t = 10 \log l_t \qquad \text{dB}. \tag{5.31}$$

The combined noise factor for two cascaded networks is (see Fig. 5.2)

$$f_{12} = f_1 + l_1(f_2 - 1). \tag{5.32}$$

Figure 5.2 Two networks in cascade.

The system noise factor f_s for the receiving system is obtained by iteratively applying (5.32) from right to left in Fig. 5.1, using (5.28) and (5.29). The result, referred to point A in Fig. 5.1, is

$$f_s = f_a + (l_c - 1)\frac{T_c}{T_o} + l_c(l_t - 1)\frac{T_t}{T_o} + l_c l_t(f_r - 1). \tag{5.33}$$

If temperatures T_c and T_t are equal to T_o, (5.33) reduces to

$$f_s = f_a - 1 + l_c l_t f_r \tag{5.34}$$

where f_s system noise factor $(F_s = 10 \log f_s)$
  f_a effective antenna noise factor
  f_r receiver noise factor.

The system noise factor f_s accounts for the contributions of external noise, receiver noise, and the losses in the antenna coupling circuit and the transmission line.

By analogy with (5.3), the system noise factor f_s is defined as the ratio of the total observable noise power at point A to the noise power available in a bandwidth b from a resistor a temperature T_o:

$$f_s = \frac{p_{RA}}{kT_ob} \tag{5.35}$$

where p_{RA} total observable noise power

kT_ob noise power in a bandwidth b from a resistor at T_o.

Then, using (5.34), we have

$$p_{RA} = kb[T_A - T_o + l_cl_tT_of_r]. \tag{5.36}$$

As a practical matter, usually either external noise or receiver noise predominates. Equations (5.33), (5.34), and (5.36) then reduce to much simpler forms. For example, below approximately 100 MHz, atmospheric and man-made noise usually exceeds receiver noise by a significant margin (assuming reasonably efficient antennas). Then (5.34) and (5.36) reduce to

$$f_s = f_a$$

and

$$p_{RA} = f_akT_ob = kT_ab.$$

Above approximately 100 MHz, external noise sources are low and receiver noise is significant. In addition, at the higher frequencies, antennas are more efficient and l_c approaches unity. Then (5.34) and (5.36) reduce to the following:

$$f_s = l_tf_r$$

and

$$p_{RA} = l_tkT_obf_r.$$

Received Noise

This method expresses receiving system performance in terms of the *received noise* referenced to the input of the receiver (equivalent input noise) and is probably more familiar to most readers.

Refer to Fig. 5.3. The noise field spectral intensity e_n or noise field power spectral density s is converted to available noise voltage v_R at the

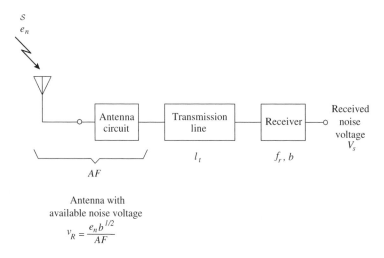

Figure 5.3   The receiving system and received noise voltage.

output of the antenna by applying the antenna factor AF. The transmission line loss factor l_t is applied, and the result is the received noise voltage v_s. Internally generated receiver noise is accounted for separately by the receiver noise factor f_r (which will be related to receiver sensitivity in the following section).

The available noise voltage v_R at the output of the antenna is (refer to Table 5.1)

$$v_R = \frac{e_n \sqrt{b}}{AF} = \frac{\sqrt{120\pi\,sb}}{AF} \qquad \text{V} \qquad (5.37)$$

where v_R available noise voltage at the antenna output, V
  e_n noise field spectral intensity, V/m/Hz$^{1/2}$
  s noise field power spectral density, W/m^2/Hz
  AF antenna factor, m^{-1}. $AF\,(\text{dB/m}) = 20\log AF$
  b   bandwidth of receiver, Hz.

When (5.37) is converted to dB and expressed in commonly used units, we have the following expressions for the available noise voltage in terms of the noise field spectral intensity and noise field power spectral density:

$$V_R(\text{dB}\mu\text{V}) = E_n(\text{dB}\mu\text{V/m/}\sqrt{\text{Hz}}) + 10\log b - AF(\text{dB/m}) \qquad (5.38)$$

and

$$V_R(\text{dB}\mu\text{V}) = \mathcal{S}(\text{dB}\mu\text{W/m}^2/\text{Hz}) + 10\log b$$
$$- AF(\text{dB/m}) + 85.76. \qquad (5.39)$$

Equating (5.38) and (5.39) results in the following expression which will be useful in converting man-made and atmospheric noise data:

$$E_n(\text{dB}\mu\text{V/m}/\sqrt{\text{Hz}}) = \mathcal{S}(\text{dB}\mu\text{W/m}^2/\text{Hz}) + 85.76. \qquad (5.40)$$

The received noise voltage v_s is just the available noise voltage at the antenna output in (5.37) reduced by the transmission line loss:

$$v_s = \frac{v_R}{\sqrt{l_t}} = \frac{e_n\sqrt{b}}{\sqrt{l_t}\,AF} = \frac{\sqrt{120\pi s b}}{\sqrt{l_t}\,AF} \qquad \text{V.} \qquad (5.41)$$

(Note: l_t was defined as the ratio of the power input to the transmission line to the power measured at the output under matched impedance conditions. The input-to-output voltage ratio is then equal to $l_t^{1/2}$.)

When (5.41) is converted to dB and expressed in the same units used in (5.38) and (5.39), we have the following expressions for the received noise voltage in terms of the noise field spectral intensity and noise field power spectral density:

$$\begin{aligned} V_s(\text{dB}\mu\text{V}) = E_n(\text{dB}\mu\text{V/m}/\sqrt{\text{Hz}}) + 10\log b \\ -AF(\text{dB/m}) - L_t \end{aligned} \qquad (5.42)$$

and

$$\begin{aligned} V_s(\text{dB}\mu\text{V}) = \mathcal{S}(\text{dB}\mu\text{W/m}^2/\text{Hz}) + 10\log b \\ -AF(\text{dB/m}) - L_t + 85.76 \end{aligned} \qquad (5.43)$$

where

$$L_t = 10\log l_t \qquad \text{cable loss, in dB.}$$

Equations (5.42) and (5.43) are used to obtain received voltage levels from published data on atmospheric and man-made noise.

Receiver Sensitivity and Noise Figure

The minimum signal level that can be detected by a receiver, a spectrum analyzer, or a field strength meter is limited by the presence of noise. In the usual case, either external noise or receiver noise is dominant. The response of a receiving system to external ambient noise sources was covered in the preceding paragraphs. The effect of receiver noise on the reception of desired signals will now be addressed.

Receiver sensitivity is defined as the input signal level that produces an output equal to twice the value of the average noise alone [2]. This definition is equivalent to a unity signal-to-noise ratio, which may be expressed as a power ratio or a voltage ratio.

Narrowband signals have a spectral occupancy which lies within the bandpass of the receiver; i.e., the signal energy is independent of bandwidth. Conversely, receiver noise power is proportional to bandwidth.

If the receiver sensitivity to narrowband signals is denoted as p_{SENS}^{NB} (input signal power which produces a unity signal-to-noise ratio) and assuming the receiver is operating in its linear region (i.e., before the detector), we have

$$\frac{p_{SENS}^{NB}}{p_s} = 1 \qquad\qquad (5.44)$$

or

$$p_{SENS}^{NB} = p_s \qquad W \qquad\qquad (5.45)$$

where p_{SENS}^{NB} narrowband receiver sensitivity
     p_s noise power of receiver (see Fig. 5.1).

Since the noise power in (5.45) is proportional to bandwidth, the narrowband receiver sensitivity is proportional to bandwidth—the greater the bandwidth, the higher (poorer) the sensitivity.

Broadband signals have a spectral occupancy which is broad compared with the bandwidth of the receiver. For impulsive broadband signals, the peak received voltage is equal to the peak spectral intensity of the signal, s, times the bandwidth of the receiver ($v_{pk} = 2V\tau b = sb$. See Chapter 6.) That is, the peak signal voltage is proportional to the bandwidth. Conversely, the rms receiver noise voltage is proportional to the square root of bandwidth.

If the receiver sensitivity to broadband signals is denoted as s_{SENS}^{BB} (input signal peak-spectral-intensity which produces a unity signal voltage-to-rms noise voltage ratio) and assuming the receiver is operating in its linear region (i.e., before the detector), we have

$$\frac{s_{SENS}^{BB} b}{V_s} = 1 \qquad\qquad (5.46)$$

or

$$s_{SENS}^{BB} = \frac{V_s}{b} \qquad V/Hz \qquad\qquad (5.47)$$

where s_{SENS}^{BB} broadband receiver sensitivity, V/Hz
     V_s rms noise voltage of receiver (see Fig. 5.3).

Since the rms noise voltage in (5.47) is proportional to the square root of bandwidth, the broadband receiver sensitivity is inversely proportional to the square root of bandwidth—the greater the bandwidth, the lower (better) the sensitivity.

Receiver noise figure F_r is a common measure of receiver noise. Noise figure F_r and noise factor f_r are related by

$$F_r = 10 \log f_r. \tag{5.48}$$

The *noise factor* of a receiver (or any two-port network) is defined as the ratio of the total noise power at the output to that portion of the output power generated by the input termination, when the input termination is at 288 K [2]. Thus, noise factor is a measure of the noise added by the receiver in excess of the noise generated by the input termination. Also note that the noise is referenced to the input.

The noise generated by the input termination is examined next. The mean-square thermal noise voltage generated by a resistor at the reference temperature $T_o = 288$ K is

$$E^2 = 4RkT_ob \tag{5.49}$$

where R is the resistance, in ohms.

Under matched-impedance conditions (receiver input impedance is equal to the termination resistance), the input noise power to the receiver is

$$p_i = \frac{(E/2)^2}{R} = kT_ob = 4 \times 10^{-21}b \qquad \text{W} \tag{5.50}$$

where p_i noise power generated by the terminating resistor, W
        $T_o = 288$ K, reference temperature
        b bandwidth, Hz.

The noise power generated by the terminating resistor, in dBm (dB referenced to one milliwatt), is

$$P_i = 10 \log p_i(\text{mW})$$

or

$$P_i = -174 + 10 \log b \qquad \text{dBm.} \tag{5.51}$$

By definition, the noise factor of the receiver is

$$f_r = \frac{p_s}{p_i} \tag{5.52}$$

where p_s total noise power of the receiver (see Fig. 5.1)
    p_i noise power generated by the terminating resistor, W.

Then

$$p_s = f_r p_i = f_r k T_o b \qquad (5.53)$$

and

$$P_s = 10 \log p_s = F_r + 10 \log p_i.$$

The total noise power at the output of the receiver in dBm, using (5.51), is

$$P_s = F_r - 174 + 10 \log b \qquad \text{dBm}. \qquad (5.54)$$

Narrowband receiver sensitivity was defined as the input signal power level which is equal to the receiver noise power level. See (5.45). We can then replace P_s with P_{SENS}^{NB} in (5.51).

The receiver sensitivity to narrowband signals in terms of the receiver noise figure and bandwidth, in dBm, is

$$\boxed{P_{\text{SENS}}^{NB} = F_r - 174 + 10 \log b \qquad \text{dBm}.} \qquad (5.55)$$

Narrowband receiver sensitivity in dB referenced to one microvolt is

$$\boxed{V_{\text{SENS}}^{NB} = F_r - 67 + 10 \log b \qquad \text{dB}\mu\text{V}.} \qquad (5.56)$$

Narrowband receiver sensitivity versus receiver noise figure from (5.55) and (5.56) is plotted in Fig. 5.4 for various bandwidths. Receiving systems with no preamplifier or preselector stage, most notably some spectrum analyzers, may have very high noise figures and resultant poor sensitivity.

Broadband receiver sensitivity was defined as the input-signal peak-spectral-intensity which is equal to the rms noise voltage of the receiver. See (5.47).

The rms noise voltage of the receiver, assuming a 50-ohm input impedance, is [see (5.53)]

$$V_s = \sqrt{50 p_s} = \sqrt{50 f_r p_i} = \sqrt{50 f_r k T_o b}. \qquad (5.57)$$

Then (5.47) becomes

$$s_{\text{SENS}}^{BB} = \frac{\sqrt{50 f_r k T_o}}{\sqrt{b}} \qquad \text{V/Hz}. \qquad (5.58)$$

In dB,

$$S_{\text{SENS}}^{BB} = 20 \log s_{\text{SENS}}^{BB}. \qquad (5.59)$$

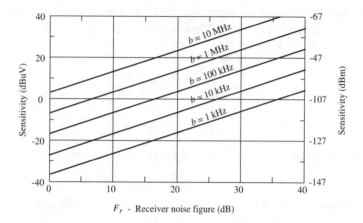

Figure 5.4 Narrowband receiver sensitivity vs. noise figure for various bandwidths.

Then,

$$S_{\text{SENS}}^{BB} = F_r - 187 - 10 \log b \qquad \text{dBV/Hz} \qquad (5.60)$$

or, in the more common units of dB referenced to one microvolt per megahertz,

$$\boxed{S_{\text{SENS}}^{BB} = F_r + 53 - 10 \log b \qquad \text{dB}\mu\text{V/MHz}.} \qquad (5.61)$$

Broadband receiver sensitivity versus receiver noise figure from (5.61) is plotted in Fig. 5.5 for various bandwidths. Note that as bandwidth increases, the broadband sensitivity improves (decreases).

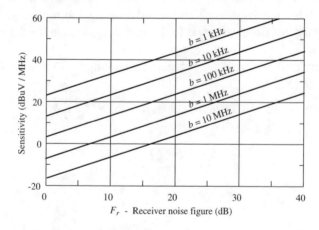

Figure 5.5 Broadband receiver sensitivity versus noise figure for various bandwidths.

The *system* noise figure F_s in (5.33) and (5.34) includes the contributions of external noise, receiver noise, and losses in the antenna circuit and transmission line. If the system noise figure is known, it may be used in place of the receiver noise figure F_r in Figs. 5.4 and 5.5 to determine the overall system sensitivity to narrowband and broadband signals.

5.3   EXTRATERRESTRIAL NOISE[1]

The primary sources of extraterrestrial or cosmic radio noise are the Milky Way galaxy and the sun. Extraterrestrial radio noise emissions are of two general types: radiation from spatially extensive sources which is broadband in nature, and radiation from spatially discrete sources which may be either narrowband or broadband.

The extraterrestrial noise spectra for a number of sources are shown in Fig. 5.6. The spectra are given in terms of both the noise field power spectral density S and the noise field spectral intensity E_n. The ordinates in Fig. 5.6 are related by (5.40) after converting the units of S from dBμW to dBW:

$$E_n(\text{dB}\mu\text{V/m}/\sqrt{\text{Hz}}) = S(\text{dBW/m}^2/\text{Hz}) + 145.8. \qquad (5.62)$$

The spatially extensive radiation consists of two components: the broadband cosmic background of magnitude 2.7 K, attributed to the residual formation energy of the universe, and the much greater galactic component, also very broadband, that arises from the multitude of unresolvable discrete sources and excited gases in our Milky Way galaxy. The magnitude of galactic noise varies somewhat with direction from the earth as a consequence of the fact that the source density is maximum in the galactic plane in the direction of the center of the galaxy. This maximum is plotted in Fig. 5.6. Noise levels from other parts of the galactic plane can be 10 to 12 dB lower. Since the galactic plane lies at an angle of approximately $60°$ to the celestial equator, the direction of the galactic background maximum is a diurnal and seasonal variable [4], [5]. Galactic noise will be put in perspective with respect to atmospheric and man-made noise in the following sections.

The discrete cosmic emission sources which may potentially contribute to received noise are the sun, the moon, and the planet Jupiter. The spectrum of each is shown in Fig. 5.6. Two sets of data are shown for the sun. The lower curve typifies the radiation from a sun manifesting a low level

---

[1] Adapted from E. N. Skomal and A. A. Smith, Jr., *Measuring the Radio Frequency Environment* [3].

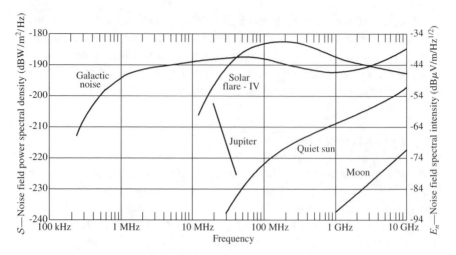

Figure 5.6 Galactic, solar, lunar, and planetary noise spectra.

of sunspot and flare activity. When solar disturbances become persistent, manifested by pronounced flare eruptions, the solar emission level increases in the frequency range below 10 GHz. Flare disturbances of Type IV, although not productive of the highest levels of radio noise, nevertheless represent those of the greatest spectral expanse with durations exceeding several days.

The rms received noise voltage at any frequency due to any of the spectra in Fig. 5.6 can be obtained by using (5.42). (See the example at the end of the next section on atmospheric noise.)

$$V_s(\text{dB}\mu\text{V}) = E_n(\text{dB}\mu\text{V/m/}\sqrt{\text{Hz}}) + 10\log b - AF(\text{dB/m}) - L_t$$

where V_s received noise voltage, $\text{dB}\mu\text{V}$
$\quad\quad E_n$ noise field spectral intensity from Fig. 5.6
$\quad\quad b$ bandwidth of the receiver, in Hz
$\quad\quad AF$ antenna factor, in dB per meter (dB/m)
$\quad\quad L_t$ cable loss, in dB.

The noise field power spectral density S in Fig. 5.6 can be converted to an effective antenna noise figure F_a if the effective aperture of the receiving antenna is known. The relation is

$$F_a = S + 10\log A_e + 204 \tag{5.63}$$

where A_e is the effective aperture in m^2. See (4.20).

5.4 ATMOSPHERIC NOISE

The source of atmospheric radio noise is lightning discharges produced during thunderstorms occurring worldwide. It is estimated that approximately 2000 thunderstorms are in progress globally at any one time, resulting in 30–100 cloud-to-ground discharges per second [6], [7]. The impulsive electromagnetic fields generated by lightning currents propagate to distant locations by ordinary ionospheric modes, while the fields from local thunderstorms arrive via groundwave propagation.

The multitude of lightning impulses arriving at any location on the earth's surface results in a nonstationary random field characterized by a lower amplitude Gaussian background with higher amplitude nonGaussian impulses superimposed [8]. In this section, only the Gaussian background of the atmospheric noise field is considered and is expressed in units related to the power spectral density (F_a, S, and E_n). The higher amplitude impulses riding on the Gaussian background have a low probability of occurrence and are described by the amplitude-probability distribution (APD) of the noise. Representative APDs can be found in [8] and [9].

Atmospheric noise varies with the time of day, the season, geographic location, and frequency. The most comprehensive summary of worldwide atmospheric noise data is contained in C.C.I.R. Report 322 [9].

Some representative atmospheric noise spectrums are given in Figs. 5.7 and 5.8. In Fig. 5.7, the atmospheric noise level is plotted in terms of the effective antenna noise factor F_a. In Fig. 5.8, the atmospheric noise is given in terms of the noise field spectral intensity E_n. Otherwise, the data presented in these two figures are identical. F_a and E_n are related by (5.16).

The spectrums in Figs. 5.7 and 5.8 are typical of atmospheric noise levels at midlatitude locations such as Chicago, New York, and Miami. Atmospheric noise is generally higher in equatorial regions and decreases with increasing latitude. Reference [9] should be consulted for noise levels elsewhere on the globe.

The summer 0000-0400 and winter 0000-0400 curves in Figs. 5.7 and 5.8 represent the highest noise levels expected during those respective seasons. Note that at 1 MHz, the winter level is approximately 20 dB lower than the summer level for the 0000-0400 time of day. The winter 0800-1200 represents the lowest atmospheric noise level that can be expected during the entire year. Above 30 MHz, atmospheric noise is negligible compared with galactic noise.

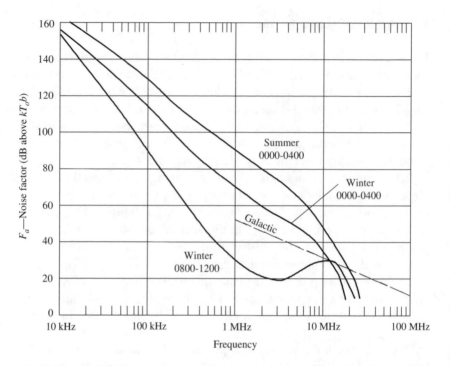

Figure 5.7 Effective antenna noise factor of atmospheric noise for midlatitude locations.

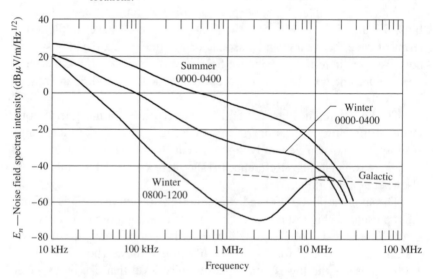

Figure 5.8 Noise field spectral intensity of atmospheric noise for midlatitude locations.

The diurnal variation of atmospheric noise at 1 MHz for a midlatitude site in summer is shown in Fig. 5.9. Note that the highest levels occur during 0000-0400 local time, while the lowest levels are observed from 0800-1200 local time. The difference between high and low atmospheric noise levels over a 24-hour period is in the order of 35 dB for the location, season, and frequency shown in Fig. 5.9.

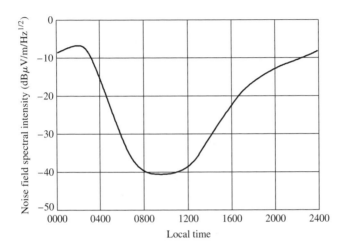

Figure 5.9 Diurnal variation of atmospheric noise spectral intensity at 1 MHz for a midlatitude site in summer.

The rms received noise voltage at any frequency due to the atmospheric noise spectra in Fig. 5.8 can be obtained by using (5.42):

$$V_s(\mathrm{dB}\mu\mathrm{V}) = E_n(\mathrm{dB}\mu\mathrm{V/m/}\sqrt{\mathrm{Hz}}) + 10\log b - AF(\mathrm{dB/m}) - L_t$$

where V_s received noise voltage, $\mathrm{dB}\mu\mathrm{V}$
 E_n atmospheric noise field spectral intensity from
       Fig. 5.8
 b bandwidth of the receiver, in Hz
 AF antenna factor, in dB per meter (dB/m)
 L_t cable loss, in dB.

EXAMPLE
Estimate the noise voltage received by a short-wave receiver (or field-strength meter or spectrum analyzer) at a frequency of 2.5 MHz due to atmospheric noise during the summer months at 2 A.M. The short-wave receiver has a bandwidth of 10 kHz and uses a remote, short vertical monopole antenna with

an antenna factor of 26 dB. The cable loss is 1 dB. The noise figure of the receiver is 10 dB.

Solution

From Figure 5.8, $E_n = -10$. Then, from the preceding equation

$$V_s = -10 + 10 \log 10^4 - 26 - 1$$

or

$$V_s = 3 \, \text{dB}\mu\text{V} \quad \text{or} \quad 1.4 \text{ microvolts.}$$

The receiver noise from Fig. 5.4 is -17 dBμV. So the received atmospheric noise level is 20 dB greater than the internal receiver noise. That is, the receiving system is atmospheric-noise limited, which is why the use of inefficient antennas is appropriate at these lower frequencies.   ∎

5.5   MAN-MADE RADIO NOISE

Man-made radio noise is broadband in nature and arises from many sources. The two major sources are power transmission and distribution lines and automotive ignition systems. Other sources include rotating electrical machinery, switching devices, appliances, light dimmers, arc generating devices, etc. Emissions from coherent sources such as clock harmonic radiation from computers, and out-of-band harmonics and spurious emissions from transmitters, are narrowband in nature. These unwanted signal interference sources are not included in the category of man-made noise.

The most comprehensive data on the amplitude and time statistics of man-made radio noise was published by Spaulding and Disney [1]. More recent measurements of business and residential noise levels made in Montreal and Ottawa, Canada, in 1993 by Lauber, Bertrand, and Bouliane [10] show that there has been no significant increase in the noise levels reported by Spaulding and Disney in [1]. In fact, the measurements in [11] "show a decrease in noise level, caused in part by the practice of using buried powerlines rather than overhead powerlines." A detailed treatment of the major sources of man-made noise can be found in Skomal's excellent book [11].

The most important parameter used to describe man-made noise is the power spectral density, which is expressed in units such as F_a, S, and E_n. Estimates of man-made radio noise for business, residential, rural, and quiet rural locations from [1], expressed in terms of the effective antenna noise factor F_a, are plotted in Fig. 5.10. The same data, converted to units of noise field spectral intensity E_n, using (5.16), are shown in Fig. 5.11.

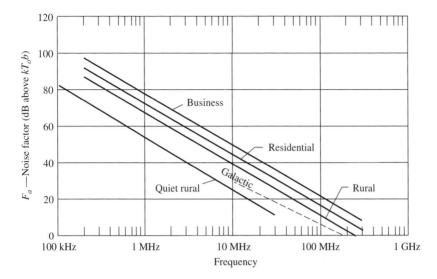

Figure 5.10 Effective antenna noise factor of man-made noise at various locations. [*After Spaulding and Disney [1]*.]

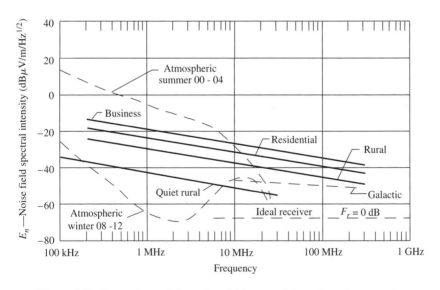

Figure 5.11 Comparison of the noise field spectral intensity of man-made, atmospheric and galactic noise.

Atmospheric noise levels, galactic noise, and receiver noise are also plotted in Fig. 5.11 for comparison purposes. The atmospheric noise curves are for the two extremes of summer 00-04 and winter 08-12 for midlatitude

locations. See Fig. 5.7. The galactic noise curve represents the maximum
level in the galactic plane in the direction of the center of the Milky Way
galaxy, as discussed in section 5.3, *Extraterrestrial Noise.* The Ideal Receiver curve is the noise voltage spectral intensity in $dB\mu V/Hz^{1/2}$ of a
receiver with a noise figure of $F_r = 0$ dB. Receiver noise levels for receivers, field-strength meters, or spectrum analyzers with a noise figure F_r
and a bandwidth b can be obtained from (5.56) or Fig. 5.4.

The rms received noise voltage at any frequency due to the noise spectra
in Fig. 5.11 can be obtained by using (5.42):

$$V_s(dB\mu V) = E_n(dB\mu V/m/\sqrt{Hz}) + 10\log b - AF(dB/m) - L_t$$

where V_s received noise voltage, $dB\mu V$
E_n noise field spectral intensity from Fig. 5.11
b bandwidth of the receiver, in Hz
AF antenna factor, in dB per meter (dB/m)
L_t cable loss, in dB.

(To calculate the received noise voltage for the Ideal Receiver curve, it is
only necessary to add $10 \log b$. The antenna factor and cable loss terms
are dropped.) Also see the example at the end of the previous section on
atmospheric noise.

The prediction of error probabilities or bit-error rates in communication
systems requires more than a knowledge of the noise power density. Refer
to [1], [8], and [10] for comprehensive data on the amplitude and time
statistics of man-made noise.

5.6   POWER-LINE CONDUCTED NOISE[2]

While the primary purpose of low-voltage power lines is the transmission
of power, they are also used as antennas for FM and TV reception and as
transmission lines for intrabuilding and interbuilding communications and
control systems. Communications and control applications include university campus radio systems, intercoms, burglar alarms, music systems,
appliance controllers, and local area networks (LANs) [12]. Noise on
power lines can affect the performance of systems that use the power lines
as antennas or transmission lines. Noise conducted on power lines can also

---

[2] Adapted from "Power Line Noise Survey" by A. A. Smith, Jr. appearing in *IEEE Transactions on Electromagnetic Compatibility*, vol. EMC-14, no. 1, pp. 31–32, February 1972.

be a source of interference if introduced into sensitive circuits of devices deriving power from the power lines. An example is interference in AM radios caused by power line conducted noise.

Measured noise spectrums on low-voltage power lines for a number of typical locations are shown in Fig. 5.12. These measurements were made at six locations ranging from a Manhattan office building to a farm in the Catskill Mountains. At each site, direct connection was made to the neutral wire at a wall outlet with a short length of center conductor from the coaxial transmission line, the outer shield being connected to conduit (green-wire) ground at the outlet. Measurements were made with a 50-ohm interference analyzer.

When the RF input of a receiving device is connected to a low-voltage power line at a wall outlet, the rms noise voltage V_s and noise power P_s (see Figs. 5.1 and 5.3) can be calculated from the noise spectra in Fig. 5.12 using the following equations:

$$V_s(\text{dB}\mu\text{V}) = V(\text{dB}\mu\text{V}/\sqrt{\text{MHz}}) + 10\log b_M - L_t \qquad (5.64)$$

$$P_s(\text{dB}\mu\text{W}) = S(\text{dB}\mu\text{W}/\text{MHz}) + 10\log b_M - L_t \qquad (5.65)$$

where V_s rms received noise voltage, dB referenced to one μV
  P_s received noise power, dB referenced to one μW
  V voltage spectral intensity (the right-hand ordinate of Fig. 5.12)
  S power spectral density (the left-hand ordinate of Fig. 5.12)
  b_M bandwidth of receiving device, in MHz
  L_t cable loss (if any), dB.

Equation (5.64) is analogous to (5.42) sans antenna factor.

V_s and P_s are related by $V_s = (50P_s)^{1/2}$ or, in dB

$$V_s(\text{dB}\mu\text{V}) = P_s(\text{dB}\mu\text{W}) + 77. \qquad (5.66)$$

The spectra in Fig. 5.12 are a measure of the Gaussian noise content of power-line noise, which is the single most important parameter. Statistics on the time-domain characteristics of the impulsive noise envelope riding on the Gaussian background are sparse. Chan and Donaldson [13] measured amplitude, width, and interarrival distributions in the frequency range below 100 kHz.

Reference [14] provides data on the attenuation of signals and noise on power distribution systems in the frequency range of 20 kHz to 30 MHz.

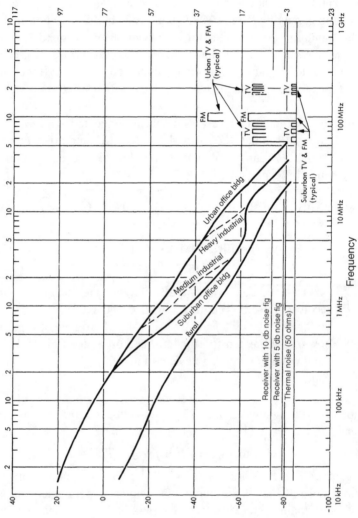

Figure 5.12 Power-line conducted noise. [©1972 IEEE.]

V - Voltage spectral intensity (dBμV/MHz^(1/2))

S - Power spectral density (dBμW/MHz)

Frequency

Urban TV & FM
(typical)

FM

TV
TV
FM
TV

Suburban TV & FM
(typical)

Urban office bldg
Heavy industrial
Medium industrial
Suburban office bldg
Rural

Receiver with 10 db noise fig
Receiver with 5 db noise fig
Thermal noise (50 ohms)

134

The paths range from those on the same branch circuit to house-to-house paths. Chan and Donaldson [15] measured the attenuation of signals on power distribution systems in the frequency range of 20 to 240 kHz.

5.7 EARTH'S MAGNETIC AND ELECTRIC FIELDS

Figure 5.13 is an idealized representation of the earth's static (DC) magnetic field. As indicated in this figure, the magnetic poles are displaced from the geographic poles (defined by the axis of rotation) by approximately 11.5° [16].

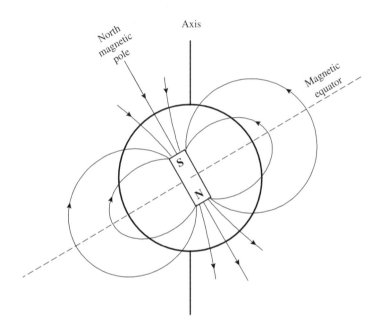

Figure 5.13 Earth's static magnetic field.

At the magnetic poles, the field is perpendicular to the surface of the earth (vertical component only). On the magnetic equator, the field is parallel to the earth's surface (horizontal component only). Anywhere else on the surface of the earth, the vector magnetic field has both a vertical and horizontal component. The field at any given location on the surface is completely specified by its magnitude and direction. The direction is designated by two angles. The angle between the magnetic field vector and the surface is called the *inclination or dip angle*. See Fig. 5.14(a). (At

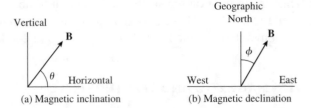

Figure 5.14 Magnetic field inclination and declination.

the magnetic poles the inclination is 90° and at the magnetic equator the inclination is 0°.) The angle between the horizontal component of the field and geographic north is called the *declination*. See Fig. 5.14(b). Global maps of equal geomagnetic flux density, inclination, and declination are published by the U.S. National Oceanographic Office. See, for instance, [17].

The magnitude of the earth's DC magnetic field at middle north-latitude locations is given in Table 5.2. The inclination at these locations is approximately 70°.

Table 5.2 Earth's Static Magnetic Field at Midlatitudes

Component	Magnitude-Gauss
B-total	0.55
B_V-vertical	0.50
B_H-horizontal	0.20

Note on polarity terminology: By definition, the north pole of a freely suspended magnet or a compass needle is the pole that points toward north. The magnetic pole at the north end of the earth is thus referred to as the north magnetic pole. However, since the north magnetic pole attracts the north pole of a compass needle, it must be the *south* pole of a magnet. That is, if the source of the earth's magnetic field is viewed as a bar magnet, the north pole of this magnet is located near the geographic south pole. See Fig. 5.13.

The earth's geomagnetic field has gone through many 180° reversals in the distant past, the last reversal occurring about 200,000 years ago. During some periods of the earth's history, reversals have occurred as frequently as every half-million years. During other periods, the time between reversals has been 50 or 60 million years. A complete reversal of the field takes approximately 5,000 to 10,000 years. Reversals of the geomagnetic field have been accompanied by simultaneous evolutionary changes in a wide variety of terrestrial life forms. Homo sapiens neandertalis emerged during the last reversal of the geomagnetic field. See the paper by Hawkins [18]. On the planet Uranus, the magnetic and geographic poles are separated by approximately 60°, suggesting that Uranus is undergoing a reversal of the north and south magnetic poles [19].

An excellent summary of the available information on the earth's electric and magnetic fields from DC to 100 Hz, including data on atmospheric noise, micropulsations, and Schumann resonances, can be found in the chapter by Polk [16]. While the earth's static electric field has not been studied as extensively as the geomagnetic field, much useful data are available. The average fair-weather electric field at the surface of the earth is approximately -130 V/m. (The average charge on the surface of the earth is negative, while the upper atmosphere carries a positive charge. See Fig. 5.15.) The fair-weather electric field decreases with altitude, as shown in Table 5.3.

The DC electric field can deviate widely from the average value and can even reverse polarity. Conditions that affect the field strength include

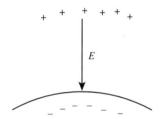

Figure 5.15 Fair-weather static E field.

Table 5.3 Earth's Fair-Weather Static Electric Field Magnitude (Vertical)

Altitude	E-V/m (average)
0	-130
100 m	-100
1 km	-45
2.4 km	-20
20 km	-1

temperature, relative humidity, fog, rain, and cloud cover. The field strength
can reach 10,000 V/m beneath thunderclouds.

REFERENCES

[1] A. D. Spaulding and R. T. Disney, "Man-Made Radio Noise, Part 1, Esti-
mates for Business, Residential, and Rural Areas," U.S. Dept. of Commerce,
Office of Telecommunications, Institute for Telecommunications Sciences,
OT Report 74-38, June 1974.

[2] IEEE Std 100-1977, *IEEE Standard Dictionary of Electrical and Electronic
Terms*, Second Edition, Institute of Electrical and Electronic Engineers, New
York, distributed in cooperation with Wiley Interscience, 1977.

[3] E. N. Skomal and A. A. Smith, Jr., *Measuring the Radio Frequency Environment*, Van Nostrand Reinhold Company, New York, 1985.

[4] H. V. Kane, "Spectra of the Non-Thermal Radio Radiation from the Galactic
Polar Regions," *Monthly Notes Royal Astron. Soc.*, vol. 189, pp. 465–478,
1979.

[5] A. Boischot, "A. Bruits Cosmiques," *NATO/AGARD Conf. Proc. Electromagnetic Noise, Interference and Compatibility*, pp. 4.1–4.12, November
1975.

[6] M. A. Uman, *Understanding Lightning*, Bek Technical Publications, Carne-
gie, PA, 1971.

[7] C. E. P. Brooks, "The Distribution of Thunderstorms Over the Globe," *Geo-
phys. Mem., London*, vol. 24, 1925.

[8] R. T. Disney and A. D. Spaulding, "Amplitude and Time Statistics of Atmospheric and Man-Made Radio Noise," ESSA Technical Report ERL 150-ITS
98, U.S. Department of Commerce, Boulder, CO, February 1970.

[9] C.C.I.R. Report 322, "World Distribution and Characteristics of Atmospheric
Radio Noise," Xth Plenary Assembly, Geneva, 1963.

[10] W. R. Lauber, J. M. Bertrand, and P. R. Bouliane, "An Update of CCIR Business and Residential Noise Levels," *1994 IEEE International Symposium on
Electromagnetic Compatibility Record*, Chicago, August 22–26, 1994.

[11] E. N. Skomal, *Man-Made Radio Noise*, Van Nostrand Reinhold Co., New
York, 1978.

[12] R. A. Piety, "Intrabuilding Data Transmission Using Power-Line Wiring,"
Hewlett-Packard Journal, May 1987.

[13] M. H. L. Chan and R. W. Donaldson, "Amplitude, Width, and Interarrival
Distributions for Noise Impulses on Intrabuilding Power Line Communication Networks," *IEEE Transactions on Electromagnetic Compatibility*, vol.
31, no. 3, pp. 320–323, August 1989.

[14] A. A. Smith, Jr., "Propagation of Interference Currents on Power Mains," *Proc. of the 3rd Symp. and Tech. Exhibition on Electromagnetic Compatibility*, Rotterdam, Netherlands, May 1–3, 1979.

[15] M. H. L. Chan and R. W. Donaldson, "Attenuation of Communication Signals on Residential and Commercial Intrabuilding Power-Distribution Circuits," *IEEE Transactions on Electromagnetic Compatibility*, vol. 28, no. 4, pp. 220–230, November 1986.

[16] C. Polk, "Sources, Propagation, Amplitude and Temporal Variation of Extremely Low Frequency (0-100 Hz) Electromagnetic Fields," Chapter II in *Biologic and Clinical Effects of Low-Frequency Magnetic and Electric Fields*, edited by J. G. LLaurado, A. Sances, Jr., and J. H. Battocletti; Charles C. Thomas, Publisher, Springfield, Illinois, 1974.

[17] *McGraw-Hill Encyclopedia of Geological Sciences*, second edition, McGraw-Hill Book Company, New York, 1988, pp. 236–243.

[18] L. H. Hawkins, "Biological Effects of Electromagnetic Fields," *Electric Field Phenomena in Biological Systems*. Proceedings of a One Day Meeting of the Static Electrification Group of the Institute of Physics, March 1989, pp. 27–38.

[19] C. Sagan, *Pale Blue Dot*, Random House, New York, 1994, p. 130.

CHAPTER 6

WAVEFORMS AND SPECTRAL ANALYSIS

This chapter begins by classifying waveforms as either energy signals or power signals. An energy signal has a continuous spectrum, while a periodic power signal has a line spectrum. The two-sided Fourier transform and related amplitude spectrum are briefly reviewed. Next, the more practical and commonly used one-sided spectral intensity is reviewed, including the definition of impulse bandwidth and examples of a rectangular pulse, an impulse and a trapezoidal pulse. Last, the Fourier series of periodic power signals is reviewed, with examples of the narrowband and broadband network response to periodic signals.

6.1 CLASSIFICATION OF SIGNALS

All practical signals can be classified as either *energy signals* or *power signals*. The distinction between energy and power is made on the basis of the following equations. Let $f(t)$ be an arbitrary signal. For example, $f(t)$ could be a voltage, a current, an electric field, or a magnetic field. The total energy W associated with $f(t)$ is

$$W = \lim_{t \to \infty} K \int_{-T}^{T} f^2(t)\, dt \qquad \text{W-sec or J} \qquad (6.1)$$

and the average power P is

$$P = \lim_{t \to \infty} \frac{K}{2T} \int_{-T}^{T} f^2(t)\, dt \qquad \text{W or J/sec} \qquad (6.2)$$

where K is a constant dependent on the particular circuit or system. If $f(t)$

141

is an electric or magnetic field, the units on (6.2) would be W/m^2, or power density.

Note that these integrals are evaluated over all time ($T \rightarrow \infty$). Two possibilities exist [1]. Either

- the total energy is finite (and thus the average power is zero)

or

- the average power is finite (and thus the total energy is infinite).

Energy signals exist only over a finite time interval and are nonperiodic. There are two classes of energy signals—*deterministic* signals and *random* signals [2]. Examples of deterministic signals include single pulses or impulses, a burst of pulses or impulses, and a sinusoidal burst (radar pulse). Deterministic energy signals have a continuous frequency spectrum, described mathematically by the Fourier transform and related transforms. Random energy signals, or noise, cannot be represented as predictable functions of time and can only be described in terms of statistical parameters.

Power signals are defined over infinite time intervals (or, realistically, time intervals that are long compared to the measurement time). There are two main classes of power signals—*periodic* signals and *random* signals [2]. Examples of periodic power signals include sine waves and pulse trains. Periodic power signals have a line spectrum of harmonically related discrete frequencies described mathematically by the Fourier series. Random power signals (noise) are characterized by various statistical parameters, the most important being power spectral density. See Section 5.1 in Chapter 5.

6.2   FOURIER TRANSFORM

The Fourier transform of a deterministic energy signal $f(t)$ is

$$F(f) = \int_{-\infty}^{\infty} f(t)\varepsilon^{-j2\pi ft}\, dt \qquad (6.3)$$

and the inverse Fourier transform is

$$f(t) = \int_{-\infty}^{\infty} F(f)\varepsilon^{j2\pi ft}\, df \qquad (6.4)$$

where f is the frequency, in Hz. In some applications, the Fourier transform is expressed in terms of the radian frequency $\omega = 2\pi f$ and is denoted as $F(\omega)$.

The Fourier transform is a complex function of frequency f and may be expressed as

$$F(f) = |F(f)|\, \varepsilon^{j\theta(f)}.$$

The magnitude $|F(f)|$ is called the *amplitude spectrum*. The square of the magnitude of the Fourier transform $|F(f)|^2$ is called the *energy density spectrum*.

The Fourier transform of a single rectangular pulse of amplitude V volts and width τ seconds is

$$F(f) = V\tau \frac{\sin \pi \tau f}{\pi \tau f} \qquad -\infty < f < \infty. \tag{6.5}$$

Fig. 6.1 depicts the Fourier transform and amplitude spectrum of a rectangular pulse.

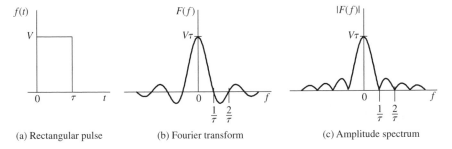

(a) Rectangular pulse (b) Fourier transform (c) Amplitude spectrum

Figure 6.1 Fourier transform and amplitude spectrum of a rectangular pulse.

The Fourier transform and amplitude spectrum are two-sided functions of frequency. This has implications when calculating the output response of a network or system with bandwidth b centered at frequency $\pm f_c$. The bandwidth must be accounted for twice since the inverse transform (6.4) extends from $-\infty$ to $+\infty$. This is illustrated in Fig. 6.2(a).

6.3 SPECTRAL INTENSITY

Spectral intensity $S(f)$ is a single-sided transform $(f \geq 0)$ related to the amplitude spectrum by

$$S(f) = 2\,|F(f)| \qquad \text{V/Hz.} \tag{6.6}$$

Negative frequencies are fictitious. Since receivers, spectrum analyzers, field strength meters, and network analyzers are tunable only over positive frequencies, spectral intensity is a more practical measure than the two-sided amplitude spectrum. Response calculations using $S(f)$ yield the same result as the amplitude spectrum since, even though the magnitude of $S(f)$ is twice as great, bandwidth is only accounted for once. See Fig. 6.2(b).

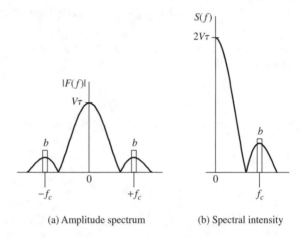

(a) Amplitude spectrum   (b) Spectral intensity

Figure 6.2 Comparison of the amplitude spectrum and the spectral intensity of a rectangular pulse at the input to a network with bandwidth b centered at frequency f_c.

NETWORK RESPONSE

The peak voltage output v_p of a unity-gain network with impulse bandwidth b centered at frequency f_c due to an input pulse with spectral intensity $S(f)$ is given by

$$v_p = S(f_c)b \qquad \text{V.} \qquad (6.7)$$

This is illustrated in Fig. 6.3. The output before and after envelope detection is shown. The waveform before detection "rings" at the center frequency f_c (which may be the IF frequency in a system using frequency conversion). The width of the output pulse is proportional to the reciprocal of the bandwidth b. It is assumed that $S(f)$ in Fig. 6.3 is constant over the bandwidth b. Impulse bandwidth is defined below.

The peak signal voltage output given by (6.7) is proportional to the bandwidth b. Conversely, rms *noise* voltage is proportional to $b^{1/2}$. Then, for broadband signals in the presence of noise, the peak signal-to-rms noise voltage ratio is proportional $b^{1/2}$. See the discussion on receiver sensitivity and noise figure in Section 5.2.

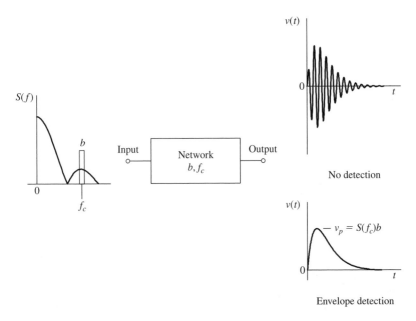

Figure 6.3 Response of a network with bandwidth b to input spectral intensity $S(f)$.

RMS Spectral Intensity

It is customary to calibrate impulse generators and field strength meters in terms of the *rms spectral intensity*, denoted $S(f)^{rms}$. This convention is a carryover from military standards, which define spectral intensity in terms of the rms value of a sine wave, whose *peak* value at the output of a network is the same as the peak response of the network to an impulse, divided by the impulse bandwidth of the network [3]. Spectral intensity $S(f)$, rms spectral intensity $S(f)^{rms}$, and the amplitude spectrum $|F(f)|$ are related by

$$S(f) = \sqrt{2}S(f)^{rms} = 2\,|F(f)| \qquad \text{V/Hz.} \qquad (6.8)$$

The peak output voltage of the network in Fig. 6.3 in terms of the rms spectral intensity is just

$$v_p = \sqrt{2}S(f_c)^{rms}b \qquad \text{V.} \qquad (6.9)$$

(Not all impulse generators and field strength meters are calibrated in terms of the rms spectral intensity, so it is important to determine which definition the manufacturer has adopted before using these instruments.)

Impulse Bandwidth Definition

Equations (6.7) and (6.9) can be taken as the defining relations for impulse bandwidth b. If v_p is the peak voltage response of a unity gain network to an impulse (see the discussion of impulses below), the impulse bandwidth is just

$$b = \frac{v_p}{S(f)} = \frac{v_p}{\sqrt{2}S(f)^{\text{rms}}} \qquad (6.10)$$

For an impulse, $S(f)$ is a constant (flat with frequency).

If the network gain is not unity, the terms on the right in (6.10) must be normalized by dividing by the gain.

Examples

Impulse. An *impulse* is defined as a pulse with a time duration that is short compared to the reciprocal of the highest frequency of interest. That is, it approximates a Dirac delta function. The shape of an impulse is not significant, only the area. The *impulse strength* A is defined as the area under the amplitude versus time curve.

The spectrum of an impulse is flat with frequency, and the spectral intensity is equal to twice the impulse strength. That is,

$$S(f) = 2A \qquad \text{V-sec or V/Hz.} \qquad (6.11)$$

An impulse and its spectral intensity are illustrated in Fig. 6.4. The spectral intensity in dBV/Hz is plotted versus log-frequency where

$$S(\text{dBV/Hz}) = 20 \log S.$$

If an impulse is passed through a unity gain network with impulse bandwidth b, the peak output voltage, from (6.7), is

$$v_p = 2Ab \qquad \text{V.} \qquad (6.12)$$

If the impulse is a narrow rectangular pulse with amplitude V and pulse width τ, the area of the impulse is $V\tau$, the spectral intensity is

$$S(f) = 2V\tau \qquad \text{V/Hz} \qquad (6.13)$$

and the peak voltage response of a unity gain network with impulse bandwidth b is

$$v_p = 2V\tau b \qquad \text{V.} \qquad (6.14)$$

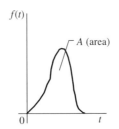

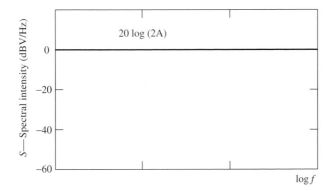

Figure 6.4   An impulse and its spectral intensity.

Rectangular Pulse.   A rectangular pulse and its spectral intensity are shown in Fig. 6.5. The amplitude of the pulse is V, and the pulse width is τ. The spectral intensity, from (6.5) and (6.6), is

$$S(f) = 2V\tau \left| \frac{\sin \pi \tau f}{\pi \tau f} \right| \qquad \text{V/Hz} \qquad (6.15)$$

where V amplitude, V
    τ pulse width, sec
    f frequency, Hz

and where

$$S(\text{dBV/Hz}) = 20 \log S.$$

The spectral intensity in dBV/Hz versus the logarithm of frequency is plotted in Fig. 6.5. In the flat region of the spectrum, the spectral intensity is a function of the area $V\tau$ and is given by

$$S(\text{dBV/Hz}) = 20 \log 2V\tau. \qquad (6.16)$$

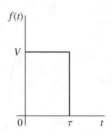

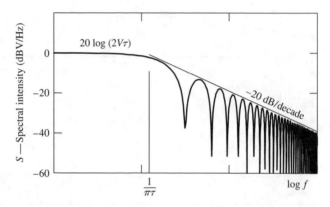

Figure 6.5 A rectangular pulse and its spectral intensity.

Beyond the flat region, the envelope of the spectrum has a slope of −20 dB per decade-frequency, or a $1/f$ fall-off. Rehkopf, in his classic paper ([4]), showed that the breakpoint between the flat and sloping regions occurs at a frequency of $1/\pi\tau$.

Trapezoidal Pulse.   Digital signals with finite rise and fall times are well represented by a trapezoidal pulse. The spectral intensity of a symmetrical trapezoidal pulse with amplitude V, pulse width is τ, and rise and fall time δ (see Fig. 6.6) is given by

$$S(f) = 2V\tau \left| \frac{\sin \pi\tau f}{\pi\tau f} \frac{\sin \pi\delta f}{\pi\delta f} \right| \qquad \text{V/Hz} \qquad (6.17)$$

where V amplitude, V
       τ pulse width, sec
       δ rise and fall time, sec
       f frequency, Hz

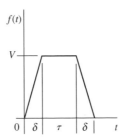

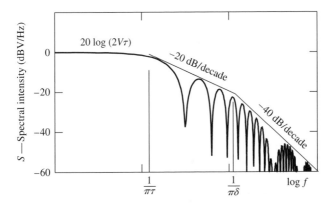

Figure 6.6 A trapezoidal pulse and its spectral intensity.

and where

$$S(\text{dBV/Hz}) = 20 \log S.$$

When the rise and fall time δ is zero, (6.17) reduces to the spectral intensity of a rectangular pulse. See (6.15). When the pulse width τ is zero, (6.17) becomes the spectral intensity of an isosceles-triangle pulse.

The spectral intensity in dBV/Hz is plotted versus log-frequency in Fig. 6.6. In the flat region of the spectrum, the spectral intensity is determined by the area of the trapezoid, the same as that of a rectangular pulse, and is given by

$$S(\text{dBV/Hz}) = 20 \log 2V\tau. \tag{6.18}$$

(It is assumed that the rise and fall times are short compared to the pulse width and thus do not contribute to the area of the trapezoid.)

Beyond the flat region of the spectrum, the slope of the spectrum envelope is initially -20 dB per decade-frequency ($1/f$ fall-off). The slope then increases to -40 dB per decade-frequency ($1/f^2$ fall-off) because of the finite rise and fall times. Rehkopf [4] showed that the breakpoint where the -40 dB per decade slope takes over occurs at a frequency of $1/\pi\delta$.

A numerical example is given in Fig. 6.7 for a trapezoidal pulse with an amplitude if 1 V, a pulse width of 1 nanosecond, and rise and fall times of 0.1 nanosecond. The spectral intensity is given in the customary units of dB referenced to one microvolt per megahertz.

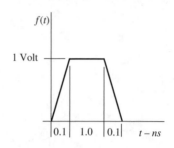

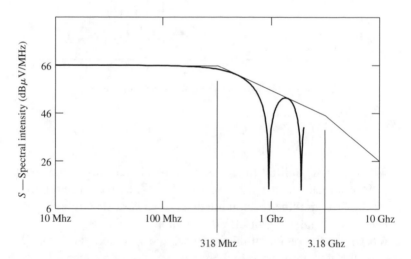

Figure 6.7 A numerical example for a trapezoidal pulse.

The spectrum of this pulse is flat at 66 dBμV/MHz out to 318 MHz ($1/\pi\tau$). The envelope of the spectrum then falls off at a rate of -20 dB per decade-frequency out to 3.18 GHz ($1/\pi\delta$), at which frequency the envelope commences a -40 dB per decade-frequency fall-off.

For a pulse amplitude V volts other than 1 volt, add

$$20\log V$$

to the ordinate in Fig. 6.7. For example, if the pulse amplitude is 10 millivolts, add

$$20 \log 0.01 = -40\text{dB}$$

to the ordinate.

6.4   FOURIER SERIES

As noted in Section 6.1, periodic signals have a line spectrum of harmonically related discrete signals described mathematically by the Fourier series. There are two forms of the Fourier series—the *exponential* form and the *trigonometric* form. The exponential form is a two-sided frequency function analogous to the Fourier transform, and has both positive and negative frequencies. The trigonometric form is a one-sided frequency function analogous to the spectral intensity discussed in the previous section. Since fictitious negative frequencies are a distraction in the real world, only the one-sided trigonometric form will be described

Let $f(t)$ be a periodic signal with period T. The trigonometric form of the Fourier series is

$$S(nf_o) = \frac{2}{T} \int_0^T f(t)\varepsilon^{-j2\pi nt/T}\, dt \qquad (6.19)$$

$$f(t) = (1/2)S_0 + \sum_{n=1}^{\infty} S(nf_o)\cos(2\pi nt/T + \phi_n) \qquad (6.20)$$

where T period of the signal
$f_o = 1/T$ fundamental frequency or pulse repetition rate
nf_o n/T spectral lines (harmonics) at integer multiples of the fundamental frequency.

Equation (6.19) is the transformation from the time domain to the frequency domain, and the Fourier coefficients $S(nf_o)$ are referred to as the *frequency spectrum* of $f(t)$. $S(nf_o)$ is analogous to the spectral intensity $S(f)$, but is defined only for the discrete frequencies $nf_{o.} = n/T$.

Equation (6.20) is the inverse transform from the frequency to the time domain. S_o is the DC component and, in the one-sided trigonometric form, is equal to twice the average value of the signal over one period.

The frequency spectrum $S(nf_o)$ is complex, having both a magnitude and phase. However, only the magnitude will be considered in the following discussion. Also, $S(nf_o)$ is the peak value of the spectral lines. The rms value of the spectral lines is

$$S(nf_0)^{rms} = S(nf_0)/\sqrt{2}. \tag{6.21}$$

The properties of the frequency spectrum $S(nf_0)$ are best illustrated by an example. Consider a train of rectangular voltage pulses with amplitude V, pulse width τ and period T. The frequency spectrum of this periodic signal is

$$S(nf_0) = \frac{2V\tau}{T} \left| \frac{\sin(n\pi\tau/T)}{n\pi\tau/T} \right| \qquad V. \tag{6.22}$$

Equation (6.22) is plotted in Fig. 6.8 for two rectangular pulse trains, both having the same amplitude V and pulse width τ. The only difference is that the period of the pulse train in Fig. 6.8(a) is twice the period of the pulse train in Fig. 6.8(b). The envelope of the spectral lines is indicated by the solid curves in both examples.

Some important properties of the frequency spectrum of periodic signals can be culled from Fig. 6.8:

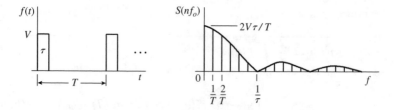

(a) Rectangular pulse train and its frequency spectrum.

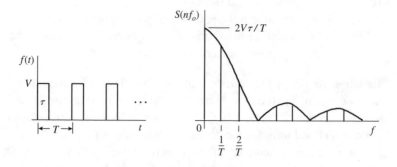

(b) Same as (a) above except that the period is reduced by one-half.

Figure 6.8 Frequency spectrum of two rectangular pulse trains.

- All the spectral lines are located at integer multiples of the fundamental frequency, that is, at frequencies of $nf_{o.} = n/T$.
- As the fundamental frequency (pulse repetition rate) f_o increases, the density of spectral lines decreases and the amplitude of the spectral lines increases.
- The *shape* of the envelope of the spectral lines is determined by the spectral intensity $S(f)$ of a single pulse.
- The *magnitude* of the spectral lines is given identically by

$$S(nf_o) = \frac{S(f)}{T} \qquad (6.23)$$

at each discrete frequency $f = nf_o = n/T$, where $S(f)$ is the spectral intensity of a single pulse. To illustrate this, replace f with n/T in (6.15) and compare with (6.22).

In order to quickly plot the frequency spectrum of a periodic signal, first plot the envelope $S(f)/T$ and then fill in the spectral lines at frequencies $f = nf_o = n/T$.

Narrowband and Broadband Response

Figure 6.9 illustrates the response of a unity gain network having a bandwidth b to a periodic signal with frequency spectrum $S(nf_o)$. The *narrowband* case is shown in Fig. 6.9(a), where the bandwidth b is "tuned" to a single spectral line at the harmonic frequency $nf_o = n/T$. When a single spectral line occupies the bandpass filter, the time domain output $f(t)$ of the filter is a sine wave of frequency nf_o that has a peak amplitude of

$$v_p = S(nf_o). \qquad (6.24)$$

The rms amplitude of the output sine wave is, from (6.21)

$$v_{rms} = S(nf_o)/\sqrt{2}. \qquad (6.25)$$

Most spectrum analyzers and field strength meters are calibrated to indicate rms values.

The *broadband* response is illustrated in Fig. 6.9(b). In this case, several spectral lines occupy the bandwidth b, and the time domain output $f(t)$ of the bandpass filter is a train of pulses. (The output is shown after envelope detection.)

The peak amplitude of the output pulses is equal to the sum of the amplitudes of the spectral lines in the bandpass, assuming that they add

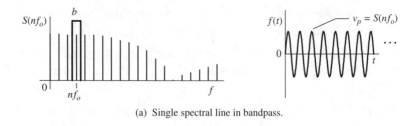

(a) Single spectral line in bandpass.

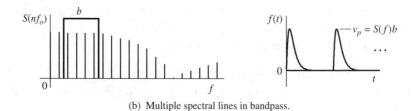

(b) Multiple spectral lines in bandpass.

Figure 6.9 Narrowband and broadband response of a bandpass filter with bandwidth b to the spectrum of a periodic signal.

in phase. (This is a practical assumption if the bandwidth occupies only a
small portion of the spectrum.) Further, assume that the frequency spectrum
is flat over the bandwidth b. The peak voltage is then

$$v_p = \sum_N S(nf_o) = N S(nf_o) \qquad (6.26)$$

where N is the number of spectral lines in the bandwidth. Recall (6.23):

$$S(nf_o) = \frac{S(f)}{T}. \qquad (6.23)$$

Substituting (6.23) into (6.26) yields

$$v_p = \frac{N S(f)}{T}. \qquad (6.27)$$

But the bandwidth b is equal to the number of spectral lines in the bandwidth
times their spacing. That is,

$$b = Nf_o = \frac{N}{T}. \qquad (6.28)$$

Then (6.27) can be written as

$$\boxed{v_p = S(f)b} \qquad (6.29)$$

where $S(f)$ is the spectral intensity of a single pulse of the periodic signal.

This is the same result obtained for the response of a single pulse in Section 6.3. See (6.7). The peak voltage response to a broadband signal is the same whether the signal is a single pulse or a train of pulses.

The width of the pulses at the output of the bandpass filter in Fig. 6.9(b) is proportional to the reciprocal of the bandwidth b. (The exact expression depends on how pulse width is defined.) That is,

$$\tau_o \propto \frac{1}{b}. \tag{6.30}$$

For example, if the bandwidth b is 1 MHz, the output pulse width is in the order of 1 microsecond.

Last, the spacing of the output pulses in Fig. 6.9(b) is equal to the period T of the periodic input signal.

REFERENCES

[1] D. K. Frederick and A. B. Carlson, *Linear Systems in Communications and Control*, John Wiley & Sons, New York, 1971.

[2] D. F. Mix, *Random Signal Analysis*, Addison-Wesley, Reading, MA, 1969.

[3] R. B. Andrews, Jr., "An Impulse Spectral Intensity Measurement System," *IEEE Transactions on Instrumentation and Measurement*, vol. IM-15, no. 4, December 1966, pp. 299–303.

[4] H. L. Rehkopf, "Prediction of Pulse Spectral Levels," *Fourth National IRE Symposium on RFI*, June 28–29, 1962.

CHAPTER 7

TRANSMISSION LINES

Two-conductor transmission line theory is reviewed in this chapter. The conditions for transverse electromagnetic (TEM) wave propagation are examined. Topics covered include the distributed parameters of transmission lines, propagation constants, characteristic impedance, and reflection and transmission coefficients. Solutions for the frequency-domain current and voltage distributions on lossy and lossless lines, including the load response, are provided. The excitation of transmission lines by external electromagnetic fields is reviewed. Radiation from common-mode and differential-mode currents on transmission lines is examined.

7.1 EXAMPLES OF TRANSMISSION LINES

Transmission lines guide electromagnetic waves from a source to a load. The cross sections of some common types of transmission lines are illustrated in Fig. 7.1. Applications span the electromagnetic spectrum from ELF to SHF and include electric power distribution, cable TV systems, antenna systems, printed circuit boards, and microwave circuits.

All the examples in Fig. 7.1 are categorized as two-conductor transmission lines except for the multiconductor line. Two-conductor lines are analyzed by the methods of classical transmission line theory, which is reviewed in the following sections.

The analysis of multiconductor transmission lines is more formidable than the analysis of two-conductor lines and requires matrix methods and notation. The reader interested in multiconductor transmission lines is referred to the definitive work on the subject by Clayton Paul. See [1].

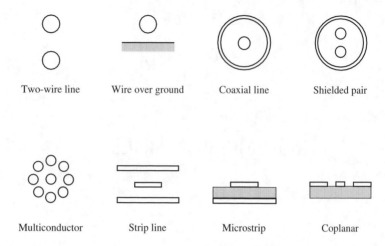

Figure 7.1 Cross sections of some common types of transmission lines.

7.2   TRANSVERSE ELECTROMAGNETIC (TEM) MODE OF PROPAGATION

Transmission line analysis is based on the basic assumption that the field surrounding the conductors is a transverse electromagnetic wave (TEM wave). A TEM wave is one in which both the electric and magnetic field vectors are perpendicular to the direction of propagation. That is, the longitudinal components of E and H are zero.

The TEM-wave assumption requires that the transmission line structure satisfy certain conditions in order for transmission line theory to be strictly applicable. Specifically,

- the line must be uniform
- the spacing between conductors must be electrically small
- the surrounding medium must be homogeneous
- the conductors must be lossless
- the currents on the line must be differential-mode currents.

These conditions are discussed below.

A *uniform line* is one in which the conductors are parallel to each other and have a uniform cross section along the line axis. Nonuniform lines have cross-sectional dimensions that vary along the line axis. Examples include tapered lines and lines with step discontinuities and gaps in the conductors.

The *spacing* between the conductors must be small compared to a wavelength. If the electrical spacing between conductors is large, generally in the order of one-half wavelength depending on the particular line geometry, higher order TE (transverse electric) and TM (transverse magnetic) modes of propagation will exist along with the TEM mode. If these *waveguide modes* exist, transmission line analysis alone will not predict the entire response. Another reason for the small- spacing requirement is to minimize radiation from differential-mode currents on unshielded lines.

Another condition for TEM-mode propagation is that the conductors must be surrounded by a *homogeneous medium*. An inhomogeneous surrounding medium might consist of several dielectrics or have a permittivity ε_r that varies with the radial or longitudinal position. This violates the TEM assumption since there is more than one velocity of propagation in the surrounding medium. If an effective dielectric constant can be calculated from the geometry and properties of the regions making up the medium, propagation on the line can be treated as propagation on an equivalent TEM line. This is referred to as a *quasi-TEM* solution. See [1] and [2].

The transmission line conductors must be *lossless* ($\sigma = \infty$). Lossy conductors invalidate the TEM assumption since currents flowing through a lossy conductor generate an electric field in the direction of propagation. If the losses are small, the TEM solution is approximately correct and is referred to as the *quasi-TEM* solution. The losses are accounted for by a distributed series resistance parameter, R.

Generally, both *differential-mode currents* and *common-mode currents* are present on a transmission line. This is illustrated in Fig. 7.2 for a two-conductor line. Transmission line theory and the TEM propagation mode assumption apply only to the differential-mode currents on the line. The differential-mode currents I_{DM} on the two conductors in Fig. 7.2 are, by definition, equal in magnitude and opposite in direction at any cross section in the line. The term differential mode is also referred to as transmission line mode, normal mode, metallic mode, bidirectional mode, and odd mode.

Common-mode currents I_{CM} on a transmission line driven by a voltage source can arise from asymmetries in the location of the source and load,

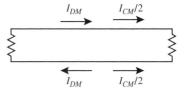

Figure 7.2 Differential-mode and common-mode currents on a two-conductor transmission line.

and from unbalances caused by the presence of nearby metal objects. See Paul [1]. Leakage through the shield of shielded cables can also produce common mode currents on the outer surface of the shield. These processes are referred to as differential-mode to common-mode conversion. The common mode current on a transmission line is distributed among the conductors. (In the case of multiconductor cables, the current is not necessarily distributed equally among the conductors since some of the interior conductors may be shielded by the outer conductors.) For the two-conductor line in Fig. 7.2, the common-mode currents on the conductors are equal in magnitude and flow in the same direction. The term common mode is also referred to as antenna mode, dipole mode, longitudinal mode, codirectional mode, and even mode.

Referring to Fig. 7.2, the response of a current probe clamped around both conductors would be

$$I_{DM} - I_{DM} + I_{CM}/2 + I_{CM}/2 = I_{CM}. \tag{7.1}$$

Only the differential-mode current flows in the terminating impedances. The common-mode current distribution is zero at both ends of the line.

Both common-mode currents and differential-mode currents can be *induced* on transmission lines by external electromagnetic fields. This subject is reviewed in Section 7.8.

Conversely, both common-mode and differential-mode currents on transmission lines are *sources* of radiated electromagnetic fields. The radiated fields produced by common-mode currents are much higher than the fields produced by like-magnitude differential-mode currents. Since differential-mode currents on the conductors flow in opposite directions, they tend to cancel; differential-mode radiation is proportional to the space-phase difference between the conductors. This topic is discussed in Section 7.9.

7.3   TWO-CONDUCTOR TRANSMISSION LINE MODEL

A two-conductor transmission line driven at one end and terminated at the other end is depicted schematically in Fig. 7.3. The conductors of the line are oriented parallel to the z-axis. The length of the line is s. The driving source at the sending end of the line has an open-circuit voltage V_G and an internal impedance Z_G. The line is terminated at the receiving end with a load impedance Z_L. $V(z)$ and $I(z)$ are the voltage and current distributions

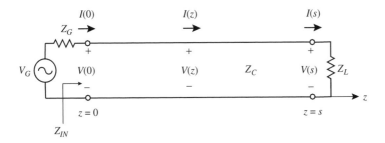

Figure 7.3 Schematic of a two-conductor transmission line with generator and
load.

on the line. $V(0)$ and $I(0)$ are the voltage and current at the sending end of
the line. $V(s)$ and $I(s)$ are the voltage and current at the receiving end, that
is, the load voltage and load current. Z_C is the characteristic impedance of
the line, and Z_{IN} is the input impedance.

The distributed parameters R, L, G, and C of a transmission line
(see Section 7.4) determine the electrical properties of the line, the most
important of which are (see Section 7.5):

- the *characteristic impedance Z_C*
- the *propagation constant γ*, consisting of
    - the *attenuation constant α*
    - and the *phase constant β*
- the *phase velocity v.*

The above electrical properties and the terminations at the ends of the
line determine the following quantities (see Sections 7.6 and 7.7):

- the reflection and transmission coefficients
- the input impedance Z_{IN}
- the voltage and current distributions along the line $V(z)$ and $I(z)$
- and the terminal voltages and currents, $V(0)$, $I(0)$, $V(s)$, and $I(s)$.

7.4   DISTRIBUTED PARAMETERS

The equivalent circuit of a differential section dz of a two-conductor transmission line is shown in Fig. 7.4. The distributed circuit parameters, or

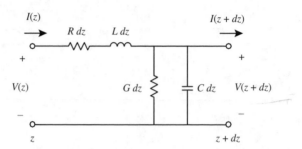

Figure 7.4 Equivalent circuit of a differential section of a two-conductor line.

per-unit-length parameters, are

R series resistance, ohms/meter
L series inductance, henrys/meter
G shunt conductance, mhos/meter
C shunt capacitance, farads/meter.

The series impedance per-unit-length is

$$Z = R + j\omega L \tag{7.2}$$

and the shunt admittance per-unit-length is

$$Y = G + j\omega C. \tag{7.3}$$

The basic differential equations for a uniform transmission line are derived from the equivalent circuit in Fig. 7.4. They are

$$\frac{d^2 V}{dz^2} - ZYV = 0 \tag{7.4}$$

and

$$\frac{d^2 I}{dz^2} - ZYI = 0. \tag{7.5}$$

Equations (7.4) and (7.5) are sometimes referred to as the *Telegrapher's equations*.

7.5   PROPAGATION CONSTANT AND CHARACTERISTIC IMPEDANCE

The solutions for the voltage and current in (7.4) and (7.5) take the following forms:

$$V = V_1 \varepsilon^{j\omega t} \varepsilon^{\gamma z} + V_2 \varepsilon^{j\omega t} \varepsilon^{-\gamma z} \tag{7.6}$$

and

$$I = \frac{V_1}{Z_C}\varepsilon^{j\omega t}\varepsilon^{\gamma z} - \frac{V_2}{Z_C}\varepsilon^{j\omega t}\varepsilon^{-\gamma z}. \tag{7.7}$$

The quantity γ is called the *propagation constant*, defined by

$$\boxed{\gamma = \alpha + j\beta = \sqrt{ZY} = \sqrt{(R + j\omega L)(G + j\omega C)}} \tag{7.8}$$

where α is the *attenuation constant*, nepers/m

β is the *phase constant*, rad/m.

The *attenuation constant in dB per meter*, denoted by $\alpha_{\text{dB/m}}$, is

$$\boxed{\alpha_{\text{dB/m}} = 8.686\alpha.} \tag{7.9}$$

(One neper is equivalent to 8.686 dB. The difference arises from the definition of the two units. The neper is equal to the natural logarithm of the ratio of two voltages or currents, while the decibel is equal to 20 times the logarithm to the base ten of the ratio of two voltages or currents.)

The quantity Z_C in (7.7) is the *characteristic impedance* of the line and is defined by

$$\boxed{Z_C = \sqrt{\frac{Z}{Y}} = \sqrt{\frac{R + j\omega L}{G + j\omega C}} \quad \text{ohms.}} \tag{7.10}$$

If $\alpha + j\beta$ is substituted for γ in (7.6) and (7.7), these equations assume the following form:

$$V = V_1\varepsilon^{\alpha z}\varepsilon^{j(\omega t + \beta z)} + V_2\varepsilon^{-\alpha z}\varepsilon^{j(\omega t - \beta z)} \tag{7.11}$$

and

$$I = \frac{V_1}{Z_C}\varepsilon^{\alpha z}\varepsilon^{j(\omega t + \beta z)} - \frac{V_2}{Z_C}\varepsilon^{-\alpha z}\varepsilon^{j(\omega t - \beta z)}. \tag{7.12}$$

The first terms in (7.11) and (7.12) are *backward-traveling waves* on the transmission line, traveling in the $-z$ direction. The second terms in (7.11) and (7.12) are *forward-traveling waves* on the line, traveling in the $+z$ direction. The terms V_1 and V_2 are complex-valued constants that are determined by the terminal conditions at the ends of the line.

The terms $\varepsilon^{\alpha z}$ and $\varepsilon^{-\alpha z}$ represent the attenuation of the backward-traveling and forward-traveling wave amplitudes, respectively. For example, the attenuation over a segment of line of length l is, by definition, the ratio of the amplitudes of the traveling wave at the beginning and the end of the segment, or

$$\frac{V(z)}{V(z+l)} = \frac{\varepsilon^{-\alpha z}}{\varepsilon^{-\alpha(z+l)}} = \varepsilon^{\alpha l} \tag{7.13}$$

where α = the attenuation constant, nepers per meter.

The attenuation in nepers is

$$\ln \varepsilon^{\alpha l} = \alpha l \qquad \text{nepers.} \tag{7.14}$$

The attenuation in dB is

$$20 \log \varepsilon^{\alpha l} = 8.686 \alpha l = \alpha_{\text{dB/m}} l \qquad \text{dB.} \tag{7.15}$$

The velocity of propagation of the waves on a transmission line, referred to as the phase velocity, is obtained from (7.11) and (7.12) by equating the phase angle $(\omega t - \beta z)$ to a constant and taking the time derivative. See Johnson [3]. The result is

$$\omega - \beta \frac{dz}{dt} = 0. \tag{7.16}$$

The phase velocity is $v = dz/dt$, or

$$\boxed{v = \frac{\omega}{\beta} \qquad \text{m/sec}} \tag{7.17}$$

where $\omega = 2\pi f$, radian frequency
   $\beta = 2\pi/\lambda$, phase constant
   $\lambda = v/f$, wavelength on the line.

Lossless Lines

A lossless transmission line is one with perfect conductors ($\sigma = \infty$) and a lossless surrounding medium (the permittivity ε is real). On a lossless line, the characteristic impedance is real and the attenuation constant is zero. Most physically short radio frequency lines can be considered lossless. On a lossless line, $R = 0$ and $G = 0$ and the series impedance per-unit-length and shunt admittance per-unit-length in (7.2) and (7.3) reduce to

$$Z = j\omega L \tag{7.18}$$

and

$$Y = j\omega C. \tag{7.19}$$

The propagation constant, (7.8), for a lossless line becomes

$$\boxed{\gamma = j\beta = j\omega\sqrt{LC}.} \tag{7.20}$$

The attenuation constant $\alpha = 0$, and the phase constant is

$$\beta = \omega\sqrt{LC} = \omega\sqrt{\mu\varepsilon} = 2\pi/\lambda \qquad \text{rad/sec} \qquad (7.21)$$

where μ permeability of the medium
ε permittivity of the medium
λ wavelength on the line.

The characteristic impedance, (7.10), for a lossless line is

$$Z_C = \sqrt{\frac{L}{C}} \qquad \text{ohms.} \qquad (7.22)$$

The *wave impedance* Z_w of the TEM wave on a transmission line (see Section 2.12 in Chapter 2) is equal to the intrinsic impedance of the medium and is given by

$$Z_W = \sqrt{\frac{\mu}{\varepsilon}}. \qquad (7.23)$$

If the surrounding medium is air, the wave impedance is that of free space:

$$Z_o = \sqrt{\frac{\mu_o}{\varepsilon_o}}. \qquad (7.24)$$

The characteristic impedance of a transmission line, defined as the ratio of the voltage between the conductors to the current flowing on the conductors, is a function of the geometry of the line *and* the medium. The wave impedance, defined as the ratio of the transverse electric and magnetic fields, is a function of the surrounding medium only. Collin [4] has shown that the characteristic and wave impedances are related by

$$Z_C = \frac{\varepsilon}{C} Z_W. \qquad (7.25)$$

The phase velocity for a lossless line, from (7.17) and (7.21), is

$$v = \frac{\omega}{\beta} = \frac{1}{\sqrt{LC}} = \frac{1}{\sqrt{\mu\varepsilon}} \qquad \text{m/sec} \qquad (7.26)$$

where μ and ε are the permeability and permittivity of the surrounding dielectric medium. (For a lossless line, the product LC depends *only* on the permeability and permittivity of the surrounding medium, and is independent of the size and spacing of the conductors.)

Lossy Lines at Radio Frequencies

At radio frequencies, $\omega L \gg R$ and $\omega C \gg G$. Depending on the particular line, these inequalities are adequately satisfied at frequencies above a few kilohertz to a few hundred kilohertz. Using these inequalities, the characteristic impedance (7.10) and phase velocity (7.17) on radio frequency lines reduce to the same form as lossless lines:

$$ Z_C = \sqrt{\frac{L}{C}} \quad \text{ohms.} \tag{7.27} $$

$$ v = \frac{1}{\sqrt{LC}} \quad \text{m/sec.} \tag{7.28} $$

At radio frequencies, Z_C and v are transmission line "constants" since L and C are independent of frequency.

The propagation constant on radio frequency lines is well approximated by

$$ \gamma = \left[\frac{R}{2Z_C} + \frac{GZ_C}{2}\right] + j\omega\sqrt{LC} \tag{7.29} $$

where the attenuation constant is

$$ \alpha(\omega) = \frac{R}{2Z_C} + \frac{GZ_C}{2} \quad \text{neper/m} \tag{7.30} $$

and the phase constant is

$$ \beta = \omega\sqrt{LC} \quad \text{rads/m.} \tag{7.31} $$

The attenuation constant $\alpha(\omega)$ increases with frequency since both distributed circuit parameters R and G increase with frequency. The series resistance R increases as the square root of frequency while G increases linearly with frequency. The first term in (7.30) is the losses in the conductors, while the second term is the dielectric losses. See Table 7.1.

Calculation of Transmission Line Constants

The electrical constants Z_C, α, β, v, and λ are functions of the per-unit-length parameters R, L, C, and G, which in turn are determined by the

Table 7.1 Design Equations for Two-Wire and Coaxial Lines at Radio Frequencies

Quantity	Two-wire line Fig. 7.5(a)	Coaxial line Fig. 7.5(b)	Units
R	$\dfrac{2}{a}\sqrt{\dfrac{f\mu}{\pi\sigma}}$	$\sqrt{\dfrac{f\mu}{4\pi\sigma}}\left(\dfrac{1}{a}+\dfrac{1}{b}\right)$	ohms/m
L	$\dfrac{\mu}{\pi}\ln\dfrac{2b}{a}$	$\dfrac{\mu}{2\pi}\ln\dfrac{b}{a}$	henrys/m
C	$\dfrac{\pi\varepsilon}{\ln(2b/a)}$	$\dfrac{2\pi\varepsilon}{\ln(b/a)}$	farads/m
G	$2\pi fC\tan\delta$	$2\pi fC\tan\delta$	mhos/m
Z_C	$\dfrac{120}{\sqrt{\varepsilon_r}}\ln\dfrac{2b}{a}$	$\dfrac{60}{\sqrt{\varepsilon_r}}\ln\dfrac{b}{a}$	ohms
α	$\dfrac{R}{2Z_C}+\dfrac{GZ_C}{2}$	$\dfrac{R}{2Z_C}+\dfrac{GZ_C}{2}$	nepers/m
v	$\dfrac{3\times10^8}{\sqrt{\varepsilon_r}}$	$\dfrac{3\times10^8}{\sqrt{\varepsilon_r}}$	m/sec

$\varepsilon=\varepsilon_o\varepsilon_r$ permittivity of dielectric medium, farads/m
$\mu=\mu_o\mu_r$ permeability of conductors, henrys/m
$\sigma=$ conductivity of conductors, mhos
$\tan\delta=$ loss tangent of dielectric medium

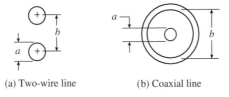

(a) Two-wire line   (b) Coaxial line

Figure 7.5 Cross-sectional dimensions of a two-wire line and a coaxial line.

geometry of the line, the conductivity and permeability of the conductors, and the permittivity of the dielectric medium.

The design equations for two-wire and coaxial lines are summarized in Table 7.1. The derivations for simple transmission line geometries can be found, for example, in the texts by Johnson [3] and Chipman [6].

Design data for more complex configurations such as shielded pairs, multiconductor cables, coplanar, strip, and microstrip lines (and their many variations), may be found in the excellent design handbook by Wadell [2], in the book by Paul [1] on multiconductor cables, and in the chapter by Itoh in [5].

In any case, the numerical calculation of transmission line parameters and constants requires a knowledge of the properties of the materials, which is the purview of designers. The user of transmission lines can usually obtain the necessary data from the manufacturers. Specifically, manufacturers publish data on the characteristic impedance, the attenuation constant, and the phase velocity (as a percentage of the speed of light in a vacuum).

7.6   REFLECTION AND TRANSMISSION COEFFICIENTS

Refer to Figure 7.3. When a wave is launched by the generator, it travels toward the load with velocity v. The time it takes this incident wave to reach the load is $T = s/v$. If the load impedance is not matched to the characteristic impedance of the line ($Z_L \neq Z_C$), part of the incident wave is reflected and travels back toward the sending end of the line. If the generator impedance is not matched to the characteristic impedance ($Z_G \neq Z_C$), the reflected wave is re-reflected and travels forward toward the load, and so on, until a state of equilibrium is reached. For a single input pulse, the waves traveling back and forth on the line attenuate and eventually die out. When the input is a sine wave, a steady-state condition is eventually reached. Solutions for the steady-state voltages and currents for sinusoidal excitations are given in the following section.

The total voltage $V(z)$ at any point on the line is the sum of the incident voltage wave $V^+(z)$ traveling in the $+z$ direction and the reflected voltage wave $V^-(z)$ traveling in the $-z$ direction. This is illustrated in Fig. 7.6. Thus,

$$V(z) = V^+(z) + V^-(z). \tag{7.32}$$

The current at any point on the line is also the sum of two waves traveling in opposite directions:

$$I(z) = I^+(z) + I^-(z). \tag{7.33}$$

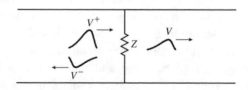

Figure 7.6 Incident, reflected, and transmitted voltages.

The incident and reflected voltages and currents are related by the characteristic impedance:

$$V^+(z) = Z_C I^+(z) \tag{7.34}$$

$$V^-(z) = -Z_C I^-(z). \tag{7.35}$$

The ratio of reflected and incident voltage waves at any point on the line is called the *voltage reflection coefficient*, denoted $\rho^V(z)$:

$$\rho^V(z) = \frac{V^+(z)}{V^-(z)}. \tag{7.36}$$

The receiving-end and sending-end *voltage reflection coefficients* are

$$\rho_L^V = \frac{V^-(s)}{V^+(s)} = \frac{Z_L - Z_C}{Z_C + Z_L} \qquad \text{receiving end} \tag{7.37}$$

$$\rho_G^V = \frac{V^-(0)}{V^+(0)} = \frac{Z_G - Z_C}{Z_C + Z_G} \qquad \text{sending end.} \tag{7.38}$$

The ratio of reflected and incident current waves at any point on the line is called the *current reflection coefficient*, denoted $\rho^I(z)$:

$$\rho^I(z) = \frac{I^+(z)}{I^-(z)}. \tag{7.39}$$

The current reflection coefficient is the negative of the voltage reflection coefficient: See (7.34) and (7.35).

The receiving-end and sending-end *current reflection coefficients* are

$$\rho_L^I = -\rho_L^V = \frac{Z_C - Z_L}{Z_C + Z_L} \qquad \text{receiving end} \tag{7.40}$$

$$\rho_G^I = -\rho_G^V = \frac{Z_C - Z_G}{Z_C + Z_G} \qquad \text{sending end.} \tag{7.41}$$

The ratio of the total voltage at an impedance (or discontinuity) to that of the incident voltage wave is called the *voltage transmission coefficient*, denoted τ^V. See Fig. 7.6. That is, at the impedance

$$\tau^V = \frac{V}{V^+} \tag{7.42}$$

or

$$V = \tau^V V^+. \tag{7.43}$$

The impedance may be a lumped element at either end of the line, or it may represent the impedance at the junction with another section of line. In the latter case, (7.43) is the voltage V transmitted beyond the junction.

The receiving-end and sending-end *voltage transmission coefficients* are

$$\tau_L^V = \frac{V(s)}{V^+(s)} = 1 + \rho_L^V = \frac{2Z_L}{Z_C + Z_L} \qquad \text{receiving end} \qquad (7.44)$$

$$\tau_G^V = \frac{V(0)}{V^-(0)} = 1 + \rho_G^V = \frac{2Z_G}{Z_C + Z_G} \qquad \text{sending end.} \qquad (7.45)$$

The *current transmission coefficients* are

$$\tau_L^I = 1 + \rho_L^I = \frac{2Z_C}{Z_C + Z_L} \qquad \text{receiving end} \qquad (7.46)$$

$$\tau_G^I = 1 + \rho_G^I = \frac{2Z_C}{Z_C + Z_G} \qquad \text{sending end.} \qquad (7.47)$$

The reflection and transmission coefficients for voltages and currents on transmission lines are analogous to the plane-wave reflection and transmission coefficients defined in Section 2.6 in Chapter 2.

7.7 SINUSOIDAL STEADY-STATE SOLUTIONS

In this section, we review the frequency-domain solutions for the voltages and currents on the two-wire transmission line illustrated in Fig. 7.3. The output of the generator V_G is a sine wave that has been applied for a sufficient time so that steady-state conditions obtain.

Voltage and Current Distributions

The voltage and current distributions on the line in Fig. 7.3, expressed in *exponential form*, are [1], [3]

$$V(z) = V_G \frac{Z_C}{Z_C + Z_G} \frac{1 + \rho_L^V \varepsilon^{-2\gamma s} \varepsilon^{2\gamma z}}{1 - \rho_L^V \rho_G^V \varepsilon^{-2\gamma s}} \varepsilon^{-\gamma z} \qquad (7.48)$$

$$I(z) = V_G \frac{1}{Z_C + Z_G} \frac{1 + \rho_L^I \varepsilon^{-2\gamma s} \varepsilon^{2\gamma z}}{1 - \rho_L^I \rho_G^I \varepsilon^{-2\gamma s}} \varepsilon^{-\gamma z} \qquad (7.49)$$

where $\rho^I = -\rho^V$.

The impedance at any point on the line is

$$Z(z) = \frac{V(z)}{I(z)} = Z_C \frac{1 + \rho_L^V \varepsilon^{-2\gamma s} \varepsilon^{2\gamma z}}{1 + \rho_L^I \varepsilon^{-2\gamma s} \varepsilon^{2\gamma z}}. \qquad (7.50)$$

Note that $Z(z)$ is a function of the characteristic impedance and the load reflection coefficient, but is independent of the sending-end reflection coefficient and impedance Z_G.

The exponential forms of the solutions are explicit functions of the propagation constant γ and the reflection coefficients of the load and source, ρ_L and ρ_G.

The voltage and current distributions on the line in Fig. 7.3, expressed in *hyperbolic notation*, are

$$V(z) = V_G \frac{Z_C Z_L \cosh \gamma(s-z) + Z_C^2 \sinh \gamma(s-z)}{(Z_C Z_G + Z_C Z_L)\cosh \gamma s + (Z_C^2 + Z_G Z_L)\sinh \gamma s} \qquad (7.51)$$

$$I(z) = V_G \frac{Z_C \cosh \gamma(s-z) + Z_L \sinh \gamma(s-z)}{(Z_C Z_G + Z_C Z_L)\cosh \gamma s + (Z_C^2 + Z_G Z_L)\sinh \gamma s} \qquad (7.52)$$

and the impedance at any point on the line is

$$Z(z) = \frac{V(z)}{I(z)} = Z_C \frac{Z_L + Z_C \tanh \gamma(s-z)}{Z_C + Z_L \tanh \gamma(s-z)}. \qquad (7.53)$$

Again, $Z(z)$ is a function of the characteristic impedance and load impedance, but independent of Z_G. It is sometimes referred to as the impedance looking toward the load, or the input impedance at point z.

Voltages and Currents at the Terminations

The *load voltage* or *receiving-end voltage* is found from (7.51) by setting $z = s$. The result is

$$V(s) = \frac{V_G Z_C Z_L}{(Z_C Z_G + Z_C Z_L)\cosh \gamma s + (Z_C^2 + Z_G Z_L)\sinh \gamma s}. \qquad (7.54)$$

The *sending-end voltage* (i.e., the voltage at the input to the line) is given by (7.51) with $z = 0$. We have

$$V(0) = V_G \frac{Z_C Z_L \cosh \gamma s + Z_C^2 \sinh \gamma s}{(Z_C Z_G + Z_C Z_L)\cosh \gamma s + (Z_C^2 + Z_G Z_L)\sinh \gamma s}. \qquad (7.55)$$

The *voltage across the sending-end impedance* Z_G is

$$V_{Z_G} = V_G - V(0) \qquad (7.56)$$

or

$$V_{Z_G} = V_G \frac{Z_C Z_G \cosh \gamma s + Z_G Z_L \sinh \gamma s}{(Z_C Z_G + Z_C Z_L)\cosh \gamma s + (Z_C^2 + Z_G Z_L)\sinh \gamma s}. \qquad (7.57)$$

The *load current* or *receiving-end current* is found from (7.52) by setting $z = s$. The result is

$$I(s) = \frac{V_G Z_C}{(Z_C Z_G + Z_C Z_L)\cosh \gamma s + (Z_C^2 + Z_G Z_L)\sinh \gamma s}. \qquad (7.58)$$

The *sending-end current* (i.e., the current flowing in the impedance Z_G) is found from (7.52) with $z = 0$. Thus,

$$I(0) = V_G \frac{Z_C \cosh \gamma s + Z_L \sinh \gamma s}{(Z_C Z_G + Z_C Z_L) \cosh \gamma s + (Z_C^2 + Z_G Z_L) \sinh \gamma s}. \qquad (7.59)$$

The *input impedance* or *sending-end impedance* Z_{IN} of the transmission line is obtained from (7.53) with $z = 0$. See Fig. 7.3. The result is

$$Z_{IN} = Z_C \frac{Z_L + Z_C \tanh \gamma s}{Z_C + Z_L \tanh \gamma s}. \qquad (7.60)$$

Matched Load.   When the load impedance is matched to the characteristic impedance of the transmission line ($Z_L = Z_C$), the load reflection coefficients are zero ($\rho_L^V = \rho_L^I = 0$) and all of the incident power is absorbed by the load.

The receiving-end (load) voltage and current, the sending-end voltage and current, and the input impedance on a line with a matched load impedance are given in Table 7.2 for both lossy and lossless lines.

For the matched load case, the exponential form of the load voltage and load current is given since this form readily conveys the physical picture of the incident wave having propagated down the length of the line.

Table 7.2 Terminal Voltages and Currents and Input Impedance for a Matched Load

Matched load	
Lossy Line	**Lossless Line**
$V(s) = \dfrac{V_G Z_L}{Z_L + Z_G} \varepsilon^{-as} \varepsilon^{-j\beta s}$	$V(s) = \dfrac{V_G Z_L}{Z_L + Z_G} \varepsilon^{-j\beta s}$
$I(s) = \dfrac{V_G}{Z_L + Z_G} \varepsilon^{-as} \varepsilon^{-j\beta s}$	$I(s) = \dfrac{V_G}{Z_L + Z_G} \varepsilon^{-j\beta s}$
$V(0) = \dfrac{V_G Z_L}{Z_L + Z_G}$	
$I(0) = \dfrac{V_G}{Z_L + Z_G}$	
$Z_{IN} = Z_L = Z_C$	

The attenuation on the line in dB, from (7.15), is

$$20 \log \varepsilon^{\alpha s} = 8.686 \alpha s = \alpha_{dB/m} s \qquad dB \qquad (7.61)$$

where s is the length of the line in meters. α is the attenuation constant in nepers/meter, and $\alpha_{dB/m}$ is the attenuation constant in dB/m. If the line is lossless, $\alpha = 0$ and $\gamma = j\beta$.

Open-Circuit Load.   When the load impedance is an open-circuit ($Z_L = \infty$), the load reflection coefficients are $\rho_L^V = +1$ and $\rho_L^I = -1$, and all of the incident power is reflected by the load.

The receiving-end (load) voltage and current, the sending-end voltage and current, and the input impedance on a line with an open-circuit load impedance are given in Table 7.3 for both lossy and lossless lines.

Table 7.3 Terminal Voltages and Currents and Input Impedance for an Open-Circuit Load

Open-circuit load	
Lossy Line	**Lossless Line**
$V(s) = \dfrac{V_G Z_C}{Z_C \cosh \gamma s + Z_G \sinh \gamma s}$	$V(s) = \dfrac{V_G Z_C}{Z_C \cos \beta s + j Z_G \sin \beta s}$
$I(s) = 0$	
$V(0) = \dfrac{V_G Z_C}{Z_C + Z_G \tanh \gamma s}$	$V(0) = \dfrac{V_G Z_C}{Z_C + j Z_G \tan \beta s}$
$I(0) = \dfrac{V_G \tanh \gamma s}{Z_C + Z_G \tanh \gamma s}$	$I(0) = \dfrac{j V_G \tan \beta s}{Z_C + j Z_G \tan \beta s}$
$Z_{IN} = Z_C \coth \gamma s$	$Z_{IN} = -j Z_C \cot \beta s$

For the open-circuit case, the hyperbolic form of the solutions is given since this form is more compact when reflections exist on the line.

Short-Circuit Load.   When the load impedance is a short-circuit ($Z_L = 0$), the load reflection coefficients are $\rho_L^V = -1$ and $\rho_L^I = +1$. As is the case with an open-circuit load, all of the incident power is reflected by the load.

The receiving-end (load) voltage and current, the sending-end voltage and current, and the input impedance on a line with a short-circuit load impedance are given in Table 7.4 for both lossy and lossless lines.

Table 7.4 Terminal Voltages and Currents and Input
Impedance for a Short-Circuit Load

Short-circuit load	
Lossy Line	**Lossless Line**
$V(s) = 0$	
$I(s) = \dfrac{V_G}{Z_G \cosh \gamma s + Z_C \sinh \gamma s}$	$I(s) = \dfrac{V_G}{Z_G \cos \beta s + jZ_C \sin \beta s}$
$V(0) = \dfrac{V_G Z_C \sinh \gamma s}{Z_G \cosh \gamma s + Z_C \sinh \gamma s}$	$V(0) = \dfrac{jV_G Z_C \sin \beta s}{Z_G \cos \beta s + jZ_C \sin \beta s}$
$I(0) = \dfrac{V_G \cosh \gamma s}{Z_G \cosh \gamma s + Z_C \sinh \gamma s}$	$I(0) = \dfrac{V_G \cos \beta s}{Z_G \cos \beta s + jZ_C \sin \beta s}$
$Z_{IN} = Z_C \tanh \gamma s$	$Z_{IN} = jZ_C \tan \beta s$

Voltage Standing-Wave Ratio (VSWR)

When the load impedance and characteristic impedance are mismatched
($Z_L \neq Z_C$), part of the incident energy is reflected at the load.   The
interference between incident and reflected waves results in *standing waves*
of voltage and current on the transmission line.

This is illustrated in Fig. 7.7. In this example, a 5-m length of 75-
ohm coaxial cable (RG-11A/U) connects a 50-ohm generator to a 50-ohm
load. The amplitude of the generator is 1 V, and the frequency is 150 MHz.

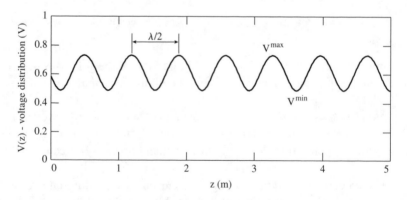

Figure 7.7 Voltage standing-wave pattern on the 5-m coaxial cable.

Referring to Fig. 7.3,

$V_G = 1$ volt	generator voltage
$s = 5$ m	length of cable
$Z_C = 75$ ohms	characteristic impedance
$Z_G = 50$ ohms	generator impedance (sending-end termination)
$Z_L = 50$ ohms	load impedance
$f = 150$ MHz	frequency.

The phase velocity of the coaxial cable is 69.5 percent of the speed of light in a vacuum. Thus

$$v = 0.695c = 0.695 \times 3 \times 10^8 \text{ m/s} = 209 \times 10^6 \text{ m/s}.$$

The wavelength on the line is

$$\lambda = \frac{v}{f} = \frac{209 \times 10^6}{150 \times 10^6} = 1.39 \text{ m}.$$

The phase constant is

$$\beta = \frac{2\pi}{\lambda} = \frac{2\pi}{1.39} = 2\pi \times 0.719 \qquad \text{rads/m}.$$

The attenuation constant for the RG-11A/U coaxial cable is 3 dB per 100 ft (0.0984 dB/m) at 150 MHz [5]. The total attenuation for the 5-m length of cable is 0.5 dB, which is negligible for purposes of this example. Thus, $\alpha \approx 0$.

Using these values, the voltage distribution $V(z)$ on the coaxial cable was calculated using (7.51) with $\alpha = 0$, $\gamma = j\beta$ (lossless case). The result is plotted in Fig. 7.7. The peaks of the voltage standing- wave are spaced one-half wavelength apart. (A similar plot of the current standing-wave would be 180° out of phase with the voltage standing-wave, i.e., the current maximums would correspond to voltage minimums.)

The *voltage standing-wave ratio* Γ is defined as the ratio of the maximum-to-minimum voltage of the standing-wave on the transmission line:

$$\Gamma = \frac{V^{\max}}{V^{\min}} \qquad \text{VSWR.} \tag{7.62}$$

The VSWR of the line in Fig. 7.7 is $\Gamma = 1.5$.

The VSWR is related to the voltage reflection coefficient of the load by [also see (4.28) to (4.30) in Section 4.1]

$$\Gamma = \frac{1 + |\rho_L^V|}{1 - |\rho_L^V|}. \tag{7.63}$$

Conversely, the receiving-end voltage reflection coefficient in terms of the VSWR is

$$|\rho_L^V| = \frac{\Gamma - 1}{\Gamma + 1}. \tag{7.64}$$

7.8   EXCITATION BY EXTERNAL ELECTROMAGNETIC FIELDS

Equations for the response of a two-wire transmission line illuminated by a nonuniform electromagnetic field were presented in the seminal paper by Taylor, Satterwhite, and Harrison [7]. A more compact form of the equations, easier to solve and allowing a clearer physical interpretation, was derived in [8]. Applications of the theory for two-conductor lines, including a conductor over a ground plane and shielded cables, can be found in the book by the author [9]. The coupling theory was extended to multiconductor lines by Paul [10]. The driving sources in all of these models are the normal component of the incident magnetic field H_y^i and the transverse component of the incident electric field E_x^i (see Fig. 7.9). The solutions are expressed in terms of the line current $I(z)$ and *total voltage* $V(z)$ on the line. This is the formulation that will be used in this section.

There are two alternative, but equivalent, forms of the solutions. In the form developed by Agrawal, Price, and Gurbaxani [11], the driving sources are the incident electric field components tangential to the conductors and to the terminations. The solutions are expressed in terms of the line current $I(z)$ and *scattered voltage* $V^s(z)$ on the line. In the formulation developed by Rachidi [12], the driving sources are the incident magnetic field components.

Those readers wishing to pursue the theory in more depth are referred to the previously mentioned book by Paul [1] and to the book by Tesche, Ianoz, and Karlsson [13]. In the book by Paul [1], general solutions for the response of two-conductor and multiconductor transmission lines to incident electromagnetic fields in both the time and frequency domains are developed. In the book by Tesche, et al., [13], a thorough theoretical analysis of field coupling to two-conductor lines, a conductor over a ground plane, and shielded cables is presented.

In the remainder of this section, the frequency-domain solutions for the load currents of a lossless two-conductor line and a conductor over a ground plane are reviewed.

Two-Conductor Line

A nonuniform electromagnetic field incident on a uniform two-conductor transmission line is depicted in Fig. 7.8. $E^i(x, y, z)$ and $H^i(x, y, z)$ are the electric and magnetic field components of the incident field. The transmission line lies in the x-z plane. The length of the line is s, the spacing between the conductors is b, and the diameter of the conductors is a [see Fig. 7.5(a)]. The characteristic impedance is Z_C. (See Table 7.1 for calculation of Z_C in terms of the line geometry and dielectric constant of the surrounding dielectric medium.) In keeping with the notation in the previous sections, Z_G and Z_L are the left-hand and right-hand load impedances, respectively. There are no lumped sources driving the line; e.g., $V_G = 0$ in Fig. 7.3. (If there are lumped sources, the total response is just the superposition of the contributions from the lumped sources and from the field-induced sources.) $I(0)$ is the current in the left-hand load, and $I(s)$

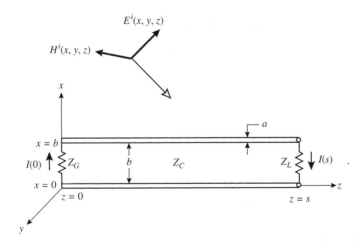

Figure 7.8 A nonuniform electromagnetic field incident on a two-wire transmission line.

is the current in the right-hand load. The load voltages are

$$V(0) = -Z_G I(0) \qquad (7.65)$$

and

$$V(s) = Z_L I(s). \qquad (7.66)$$

Figure 7.9 illustrates the components of the incident field that couple to, or excite, the transmission line. These incident field components induce distributed voltage sources along the conductors and along the terminations. The longitudinal (z-directed) electric field components incident on the two conductors, $E_z^i(b, z)$ and $E_z^i(0, z)$, excite both a *common-mode* and a *differential-mode* current on the line. See Section 7.2 and Fig. 7.2. The source of the common-mode current is the sum of the longitudinal E-fields on the two conductors, $E_z^i(b, z) + E_z^i(0, z)$. Common-mode or antenna-mode currents must be calculated using linear antenna theory. Tesche [13] calculated the common-mode and differential-mode current distributions on a 30-m long line. The common-mode current at the center of the line was larger than the differential-mode current by a factor of 20 (26 dB). The common-mode current distribution is zero at both ends of the line; common-mode currents do not flow in the load impedances.

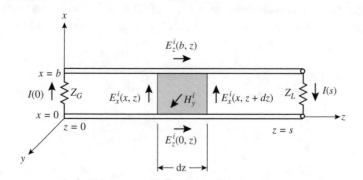

Figure 7.9 Components of the incident field that couple to the line.

The sources of differential-mode currents on the transmission line are the difference of the longitudinal E fields on the two conductors, $E_z^i(b, z) - E_z^i(0, z)$, and the transverse ($x$-directed) E-fields incident on the terminations, $E_x^i(x, 0)$ and $E_x^i(x, s)$. Only differential-mode currents flow in the loads and only differential-mode currents are predicted by transmission line theory.

Nonuniform-Field Excitation.   The load currents on a lossless two-conductor line illuminated by a nonuniform electromagnetic field are [8], [9]

$$
I(0) = \frac{1}{D} \int_0^s K(z)[Z_C \cos \beta(s-z) + jZ_L \sin \beta(s-z)]\,dz
$$

$$
+ \frac{1}{D}[Z_C \cos \beta s + jZ_L \sin \beta s] \int_0^b E_x^i(x,0)\,dx \qquad (7.67)
$$

$$
- \frac{Z_C}{D} \int_0^b E_x^i(x,s)\,dx
$$

and

$$
I(s) = \frac{1}{D} \int_0^s K(z)[Z_C \cos \beta z + jZ_G \sin \beta z]\,dz
$$

$$
- \frac{1}{D}[Z_C \cos \beta s + jZ_G \sin \beta s] \int_0^b E_x^i(x,s)\,dx \qquad (7.68)
$$

$$
+ \frac{Z_C}{D} \int_0^b E_x^i(x,0)\,dx
$$

where the denominator D in (7.67) and (7.68) is

$$
D = (Z_C Z_G + Z_C Z_L) \cos \beta s + j(Z_C^2 + Z_G Z_L) \sin \beta s \qquad (7.69)
$$

and where $K(z) = E_z^i(b,z) - E_z^i(0,z)$
$\beta = \omega/v = 2\pi/\lambda$ phase constant of line
v = phase velocity
$\omega = 2\pi f$
f = frequency, Hz.

Note: The load currents on a lossy line can be obtained from (7.67) to (7.69) by substituting $\cosh \gamma q$ for $\cos \beta q$ and $\sinh \gamma q$ for $j \sin \beta q$, where $q = z$ or s, as appropriate.

Plane-Wave Excitation.   The expressions for the load currents excited by nonuniform fields in (7.67) and (7.68) reduce to much simpler forms when the illuminating field is a plane wave. Three cases of practical interest serve to illustrate the theory: end-fire incidence, broadside incidence, and edge-fire incidence. These solutions are also derived by Paul [1] and in [9].

Case 1: End-Fire Incidence. In this case, the incident electric field E_x^i is parallel to the terminations and is traveling in the z direction. See Fig. 7.10. The incident magnetic field H_y^i is normal to the plane of the line. $E_z^i = 0$, and thus $K(z) = 0$. Also, since the electrical spacing of the conductors is small ($\beta b \ll 1$), E_x^i is uniform (constant) over the terminations and

$$\int_0^b E_x^i \, dx = b E_x^i \tag{7.70}$$

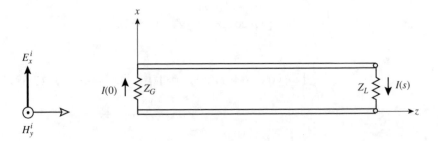

Figure 7.10 End-fire incidence.

With these simplifying conditions, the expressions for the load currents for end-fire incidence are

$$I(0) = j \frac{b E_x^i}{D} (Z_C + Z_L) \sin \beta s \tag{7.71}$$

and

$$I(s) = \frac{b E_x^i}{2D} (Z_C + Z_G)[(1 - \cos 2\beta s) + j \sin 2\beta s]. \tag{7.72}$$

Case 2: Broadside Incidence. In the case of broadside incidence shown in Fig. 7.11, the incident electric field E_x^i is parallel to the termina-tions and is traveling in the $-y$ direction. The incident magnetic field H_z^i is parallel to the plane of the line. That is, no magnetic flux lines link the plane of the transmission line when viewed as a rectangular loop. As in the case of end-fire incidence, $E_z^i = 0$ and thus $K(z) = 0$. And since E_x^i is uniform (constant) over the terminations, (7.70) holds.

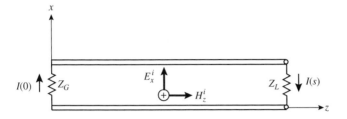

Figure 7.11 Broadside incidence.

The load currents for broadside incidence are

$$I(0) = -\frac{bE_x^i}{D}[Z_C(1 - \cos \beta s) - jZ_L \sin \beta s] \qquad (7.73)$$

and

$$I(s) = \frac{bE_x^i}{D}[Z_C(1 - \cos \beta s) - jZ_G \sin \beta s]. \qquad (7.74)$$

Note the symmetry in the expressions for the load currents.

Case 3: Edge-Fire Incidence.   In this case, shown in Fig. 7.12, the incident electric field E_z^i is parallel to the conductors and is traveling in the $-x$ direction. The incident magnetic field H_y^i is normal to the plane of the line. Also, $E_x^i = 0$ and the terminations are not illuminated.

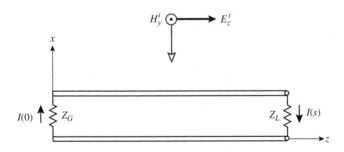

Figure 7.12 Edge-fire Incidence.

If the phase reference for the incident field is taken midway between the conductors at $x = b/2$, we have

$$E_z^i(b) = E_z^i \varepsilon^{+j\beta b/2}$$

and

$$E_z^i(0) = E_z^i \varepsilon^{-j\beta b/2}.$$

Then

$$K(z) = E_z^i \left[j2 \sin \frac{\beta b}{2} \right]. \tag{7.75}$$

The load currents for edge-fire incidence are given below. Again, note the symmetry in these expressions.

$$I(0) = -\frac{bE_z^i}{D} \left[\frac{\sin \frac{\beta b}{2}}{\frac{\beta b}{2}} \right] [Z_L(1 - \cos \beta s) - j Z_C \sin \beta s] \tag{7.76}$$

and

$$I(s) = -\frac{bE_z^i}{D} \left[\frac{\sin \frac{\beta b}{2}}{\frac{\beta b}{2}} \right] [Z_G(1 - \cos \beta s) - j Z_C \sin \beta s]. \tag{7.77}$$

Examples of load current spectrums on two-conductor lines of various lengths can be found in [1] and [9]. The results for an isolated two-conductor transmission line have limited application since this type of line is rarely used in practice. However, the two-conductor model is the basis for applications involving a conductor over a ground plane and related applications such as calculating the currents on the shield of coaxial and other shielded cables. These applications occur much more frequently in practice than applications involving two- conductor lines.

Conductor Over a Ground Plane

A single-conductor transmission line terminated at both ends to a perfectly conducting ground plane of infinite extent is illustrated in Fig. 7.13. The length of the line is s. The height of the conductor above the ground plane is h, and the diameter of the conductor is a. The terminating impedances are Z_G and Z_L.

The characteristic impedance of the line is

$$Z_C = \frac{60}{\sqrt{\varepsilon_r}} \ln \frac{4h}{a} \tag{7.78}$$

which is one-half the characteristic impedance of a two-conductor line with the conductors separated by $b = 2h$. ($2h$ is the spacing between the conductor and its image in the ground plane). See Table 7.1.

The total fields which illuminate the line are the sum of the incident (direct) fields and the ground-reflected fields. See Section 3.3, *Propagation*

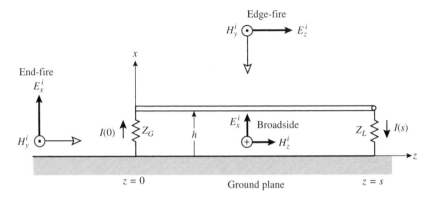

Figure 7.13 Single conductor over a ground plane.

over a Perfectly Conducting Plane. In the case of vertical polarization (end-fire and broadside incidence in Fig. 7.13), the wave reflection coefficient for a perfectly conducting plane is $\rho_v = +1$, and the total field is twice the incident field. That is,

$$E_x^{\text{total}} = 2E_x^i. \tag{7.79}$$

For horizontally polarized waves (edge-fire incidence in Fig. 7.13), the wave reflection coefficient at the perfectly conducting plane is $\rho_h = -1$. The total field on the conductor, taking the phase reference at the surface of the ground plane ($x = 0$), is

$$E_z^{\text{total}} = E_z^i \varepsilon^{+j\beta h} - E_z^i \varepsilon^{-j\beta h} \tag{7.80}$$

or

$$E_z^{\text{total}} = E_z^i [j2 \sin \beta h]. \tag{7.81}$$

Compare (7.81) with (7.75).

The solutions for the load currents for a single conductor above a ground plane can be obtained from the results for the two-conductor line by replacing b with $2h$. The solutions are given below for plane-wave excitation—specifically, for end-fire, broadside, and edge-fire incidence.

Common-mode or *antenna-mode* currents do not exist on a conductor over a ground plane. The *differential-mode* or *transmission-line-mode* current is the total current on the line.

Case 1: End-Fire Incidence.   The expressions for the load currents for end-fire incidence are (see Fig. 7.13)

$$I(0) = j\frac{2hE_x^i}{D}(Z_C + Z_L) \sin \beta s \tag{7.82}$$

and

$$I(s) = \frac{hE_x^i}{D}(Z_C + Z_G)[(1 - \cos 2\beta s) + j\sin 2\beta s]. \qquad (7.83)$$

As a practical matter, the incident and ground-reflected fields near a conducting ground plane cannot easily be separated. Only the total fields can be measured. E_x^{total} can be measured with a vertical monopole, and H_y^{total} can be measured with a loop antenna.

The load currents can be expressed in terms of the measured total electric field by making the following substitution in (7.82) and (7.83):

$$E_x^i = E_x^{\text{total}}/2. \qquad (7.84)$$

The load currents can also be expressed in terms of the measured total magnetic field. Since the incident wave is plane wave, and assuming that the surrounding medium is air $(Zo = 120\pi)$, we have

$$E_x^i = 120\pi H_y^i.$$

Also,

$$H_y^{\text{total}} = 2H_y^i.$$

We can then make the following substitution in (7.82) and (7.83):

$$E_x^i = 60\pi H_y^{\text{total}}. \qquad (7.85)$$

Case 2: Broadside Incidence.   Refer to Fig. 7.13. The load currents for broadside incidence, derived from the solutions for a two-conductor line, are

$$I(0) = -\frac{2hE_x^i}{D}[Z_C(1 - \cos \beta s) - jZ_L \sin \beta s] \qquad (7.86)$$

and

$$I(s) = \frac{2hE_x^i}{D}[Z_C(1 - \cos \beta s) - jZ_G \sin \beta s]. \qquad (7.87)$$

As in the case of end-fire incidence, the load currents may be expressed in terms of the measured total electric field by substituting (7.84) in (7.86) and (7.87).

The load currents may also be expressed in terms of the measured total magnetic field by making the following substitution equations in (7.86) and (7.87):

$$E_x^i = 60\pi H_z^{\text{total}}. \qquad (7.88)$$

(This substitution is simply a way of expressing the incident electric field in terms of the measured total magnetic field H_z^{total}, and does not imply that the magnetic flux links the plane of the line and contributes to the excitation of the line.)

Case 3: Edge-Fire Incidence.   The load currents for edge-fire incidence are (see Fig. 7.13)

$$I(0) = -\frac{2hE_z^i}{D}\left[\frac{\sin \beta h}{\beta h}\right][Z_L(1 - \cos \beta s) - jZ_C \sin \beta s] \qquad (7.89)$$

and

$$I(s) = -\frac{2hE_z^i}{D}\left[\frac{\sin \beta h}{\beta h}\right][Z_G(1 - \cos \beta s) - jZ_C \sin \beta s]. \qquad (7.90)$$

As discussed previously, it is generally not possible to measure the *incident* field near a ground plane. In the case of a horizontally polarized field, it is also difficult to measure the *total electric field*, since this field is very small and has a strong dependence on position above the ground plane. (E_z is zero on the surface of the ground plane and increases rapidly with height above the plane.) On the other hand, the total magnetic field strength near the surface of the ground is twice the incident magnetic field strength and does not vary with height above the ground plane for heights small compared to a wavelength ($\beta h \ll 1$).

The load currents for edge-fire incidence may be expressed in terms of the measured total magnetic field by making the following substitution in (7.89) and (7.90):

$$E_z^i = 60\pi H_y^{\text{total}}. \qquad (7.91)$$

Example 1: Short Conductor Over a Ground Plane.   The load voltage spectrum of a physically short conductor over a ground plane normalized to the total magnetic field for edge-fire incidence is plotted in Fig. 7.14. This example might represent an interconnecting wire routed over a circuit board in an electronic enclosure, exposed to noise fields from external or internal sources. The dimensions of the line are

$$s = 0.25 \text{ m} \qquad \text{line length}$$
$$h = 0.01 \text{ m} \qquad \text{height of conductor above ground}$$
$$a = 0.001 \text{ m} \qquad \text{diameter of conductor.}$$

The characteristic impedance from (7.78) is 221.3 ohms, and the line is matched at both ends, i.e., $Z_G = Z_L = Z_C = 221.3$ ohms.

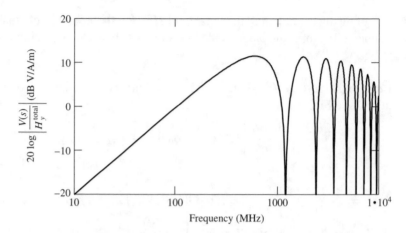

Figure 7.14 Load voltage spectrum for a short conductor over a ground plane in terms of the total magnetic field (horizontal polarization, edge-fire incidence).

The load voltage spectrum was calculated from (7.90) and (7.66), using (7.91) to express the response in terms of the total magnetic field. The resulting expression is

$$\frac{V(s)}{H_y^{\text{total}}} = -60\pi h \left[\frac{\sin \beta h}{\beta h} \right] [(1 - \cos \beta s) - j \sin \beta s]. \tag{7.92}$$

The magnitude of (7.92) in dB is plotted in Fig. 7.14.

ELECTRICALLY SHORT LINE.   When the transmission line is short compared to a wavelength ($\beta s \ll 1$), it can be approximated as a small rectangular loop antenna. The transmission line equations can be dispensed with and the load response can be calculated directly from Faraday's law. The open-circuit voltage of a loop is given by (2.3) in Chapter 2.

In the present example, the sending-end and receiving-end impedances are equal ($Z_L = Z_G$), and the voltage across each termination is one-half the open-circuit voltage.

From (2.3) in Chapter 2, the magnitude of the load voltage normalized to the total magnetic field is

$$\left| \frac{V(s)}{H_y^{\text{total}}} \right| = \frac{\omega \mu_o S}{2}. \tag{7.93}$$

In this example, the area of the line is

$$S = h \times s = 0.01 \times 0.25 = 0.0025 \text{ m}^2$$

and (7.93) reduces to

$$\left| \frac{V(s)}{H_y^{\text{total}}} \right| = 0.0098696 f_M \qquad (7.94)$$

where f_M is the frequency in megahertz. This solution is much simpler than the transmission line formulation given by (7.92).
Equation (7.94) expressed in dB is

$$20 \log \left| \frac{V(s)}{H_y^{\text{total}}} \right| = -40.1 + 20 \log f_M. \qquad (7.95)$$

The validity of the rectangular loop approximation for electrically short lines can be examined by comparing the numerical results of (7.95) with the load voltage spectrum in Fig. 7.14, which was calculated from the transmission line formulation in (7.92). This comparison shows that the rectangular loop approximation in (7.95) is accurate to within 0.6 dB when the electrical length of the line is one-quarter wavelength (240 MHz in this example). When the electrical length of the line is 0.1λ (120 MHz), the agreement is within 0.16 dB.

Example 2: Long Conductor Over a Ground Plane.   The load current spectrum of a long conductor over a ground plane normalized to the total vertical electric field for end-fire incidence is plotted in Fig. 7.15. This example might represent an outdoor overhead wire or cable exposed to

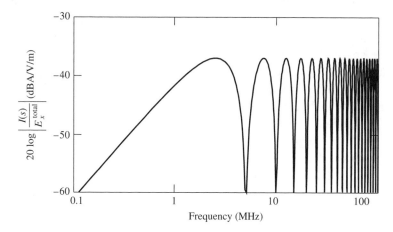

Figure 7.15 Load current spectrum for a long conductor over a ground plane in terms of the total vertical electric field (end-fire incidence).

ambient noise fields. The dimensions of the line are

$s = 30$ m   line length
$h = 3$ m   height of conductor above ground
$a = 0.01$ m   diameter of conductor.

The characteristic impedance from (7.78) is 425 ohms. The line is short circuited at the receiving end and matched at the sending end, i.e., $Z_L = 0$ and $Z_G = Z_C = 425$.

The spectrum in Fig. 7.15 was calculated from (7.83) using (7.84) to express the results in terms of the total vertical electric field.

7.9 RADIATION FROM TRANSMISSION LINES

Just as external electromagnetic fields can *induce* currents on transmission lines, the reciprocity principle dictates that currents on transmission lines will give rise to *radiated* electromagnetic fields. In general, both differential-mode and common-mode currents can exist on an isolated two-conductor or multiconductor line. In the case of a single conductor over a ground plane, the common-mode current is nonexistent; radiation is due solely to the differential-mode current. The currents on a conductor over a ground plane might represent leakage currents on the shield of a coaxial or other shielded cable.

Differential-mode or transmission line-mode currents do not radiate very efficiently. Since the differential-mode currents are equal in magnitude and oppositely directed (see Fig. 7.2), they tend to cancel; radiation is due to the space-phase difference between the conductors. Common-mode or antenna-mode currents on the conductors are equal in magnitude and flow in the same direction, resulting in much higher radiated field levels compared to differential-mode currents, as will be demonstrated below.

The radiated fields due to an arbitrary current distribution must be evaluated by integrating the current over the length of the line and the terminations. Besides the challenge of trying to accurately determine or specify the magnitude and phase of the current distribution, the resulting mathematics can be quite complex. However, if the transmission line is electrically short (approximately one-quarter wavelength or shorter), it can be modeled as a small loop antenna for differential-mode radiation, and as a pair of short dipoles for common- mode radiation. The analysis in this section assumes that the lines are electrically short.

Two-Conductor Line

Differential-Mode Radiation.   The radiated fields due to differential-mode currents on an electrically short two-conductor transmission line, modeled as a small rectangular loop antenna, are illustrated in Fig. 7.16. As previously, the length of the line is s and the spacing between the conductors is b. The area of the transmission line (rectangular loop) is

$$S = bs \qquad \text{m}^2. \tag{7.96}$$

I_{DM} is the normal signal current on the line due to sources at either end of the line and can be predicted from straightforward transmission line theory. See Section 7.7. The differential-mode current can also be measured with a current probe. This measurement may be conveniently performed at either termination where the common-mode current is zero. That is, since the line is assumed to be electrically short ($s/\lambda \ll 1$), the differential-mode current is uniform in amplitude and phase (constant) everywhere on the line.

In Fig. 7.16, r is the distance from the transmission line to the field point. It is assumed that r is much greater than the length of the line ($r \gg s$). The radiated fields are obtained directly from the solutions for

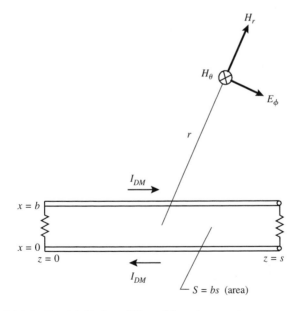

Figure 7.16 Radiated fields from differential-mode currents on a two-conductor line.

a small loop given in Section 2.10. The coordinate system is the same as in Fig. 2.15. We restrict the solutions to the fields in the same plane as the transmission line, i.e., $\theta = 90°$ in Fig. 2.15.

FAR FIELDS.   The far-field transverse electric field component, from (2.72) is

$$E_\phi = \frac{30 I_{DM} S\beta^2}{r} \varepsilon^{-j\beta r} \qquad \text{V/m} \qquad (7.97)$$

or

$$E_\phi = 1.316 \times 10^{-2} \frac{I_{DM} S f_M^2}{r} \varepsilon^{-j\beta r} \qquad \text{V/m} \qquad (7.98)$$

where f_M is the frequency in MHz and S is the area of the line. This is the same result obtained by Paul and Bush [14] and Paul [15], who modeled the conductors as infinitesimal dipoles.

The far-field transverse magnetic field component, from (2.73) is

$$H_\theta = \frac{I_{DM} S\beta^2}{4\pi r} \varepsilon^{-j\beta r} \qquad \text{A/m} \qquad (7.99)$$

or

$$H_\theta = 3.5 \times 10^{-5} \frac{I_{DM} S f_M^2}{r} \varepsilon^{-j\beta r} \qquad \text{A/m.} \qquad (7.100)$$

REACTIVE NEAR FIELDS.   The reactive near-field components due to the differential-mode current, from (2.72) to (2.74), are

$$E_\phi = -j0.628 \frac{I_{DM} S f_M}{r^2} \varepsilon^{-j\beta r} \qquad \text{V/m} \qquad (7.101)$$

$$H_\theta = \frac{I_{DM} S}{4\pi r^3} \varepsilon^{-j\beta r} \qquad \text{A/m} \qquad (7.102)$$

and

$$H_r = \frac{I_{DM} S}{2\pi r^3} \varepsilon^{-j\beta r} \qquad \text{A/m.} \qquad (7.103)$$

Common-Mode Radiation.   The radiated fields due to common-mode currents on an electrically short two-conductor transmission line are illustrated in Fig. 7.17. The conductors are modeled as two short dipole antennas spaced a distance b apart. As previously, the length of the line (short dipoles) is s. A triangular current distribution is assumed on the

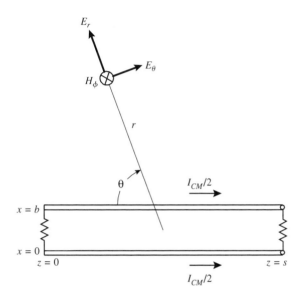

Figure 7.17 Radiated fields from common-mode currents on a two-conductor transmission.

short dipoles, which is representative of common-mode currents in that the current is zero at the ends of the line. The current at the midpoint of each conductor is $I_{CM}/2$. A current probe clamped around both conductors at the midpoint of the line would read I_{CM}.

In Fig. 7.17, r is the distance from the transmission line to the field point. It is assumed that r is much greater than the length of the line ($r \gg s$). The radiated fields are obtained from the solutions for a short dipole given by (2.68) to (2.70), Section 2.9. (These equations must be divided by 2 since a triangular current distribution is assumed in the present application.) The coordinate system is the same as in Fig. 2.12. Again, we restrict the solutions to the fields in the same plane as the line.

FAR FIELDS.   The far-field transverse electric field component, from (2.68), is

$$E_\theta = j0.3125 \frac{I_{CM} s f_M \sin\theta}{r} \varepsilon^{-j\beta r} \qquad \text{V/m} \qquad (7.104)$$

where f_M is the frequency in MHz and s is the length of the line. This result is lower than that obtained by Paul and Bush [14] and Paul [15] by a factor of 2. These investigators modeled the conductors as infinitesimal dipoles with a uniform current distribution.

The far-field transverse magnetic field component, from (2.70) is

$$H_\phi = j8.35 \times 10^{-4} \frac{I_{CM} s f_M \sin\theta}{r} \varepsilon^{-j\beta r} \qquad \text{A/m.} \qquad (7.105)$$

REACTIVE NEAR FIELDS.   The reactive near-field components due to the common-mode current, from (2.68) to (2.70), are

$$E_\theta = -j716 \frac{I_{CM} s \sin\theta}{f_M r^3} \varepsilon^{-j\beta r} \qquad \text{V/m} \qquad (7.106)$$

$$E_r = -j1432 \frac{I_{CM} s \cos\theta}{f_M r^3} \varepsilon^{-j\beta r} \qquad \text{V/m} \qquad (7.107)$$

and

$$H_\phi = 0.0398 \frac{I_{CM} s \sin\theta}{r^2} \varepsilon^{-j\beta r} \qquad \text{A/m.} \qquad (7.108)$$

COMPARISON OF DM AND CM RADIATION.   In order to compare the relative magnitudes of the radiated fields due to differential-mode and common-mode currents, consider a two-conductor line with the following dimensions (see Figs. 7.16 and 7.17):

$s = 0.25$ m   line length

$b = 0.01$ m   conductor spacing

$S = b \times s = 2.5 \times 10^{-3}$ m^2 area.

The far-field transverse E field due to differential-mode currents, from (7.98), is

$$|E_{DM}| = \frac{3.29 \times 10^{-5} I_{DM} f_M^2}{r}. \qquad (7.109)$$

The far-field transverse E field due to common-mode currents, from (7.104) with $\theta = 90°$, is

$$|E_{CM}| = \frac{7.8 \times 10^{-2} I_{CM} f_M}{r}. \qquad (7.110)$$

The ratio of the fields radiated by common-mode and differential-mode currents is

$$\frac{|E_{CM}|}{|E_{DM}|} = \frac{2.37 \times 10^3}{f_M} \frac{I_{CM}}{I_{DM}}. \qquad (7.111)$$

If the common-mode and differential-mode currents are equal, at a frequency of 100 MHz, for example,

$$\frac{|E_{CM}|}{|E_{DM}|} = 23.7 \quad \text{or} \quad 27.5 \text{ dB}. \tag{7.112}$$

That is, the radiated E field due to the common-mode current is 27.5 dB greater than the radiated E field due to the differential-mode current. This result is consistent with the measurements reported by Paul and Bush [14], [15]. While it is unlikely that the common-mode and differential-mode currents would be equal in a practical application, this analysis at least shows that common-mode currents will, most likely, be the predominant source of radiation.

Conductor over a Ground Plane

A single-conductor transmission line terminated at both ends to a perfectly conducting ground plane is shown in Fig. 7.18. The length of the line is s and the height of the conductor above the ground plane is h. The current on the conductor is I_{DM}. The current distribution is uniform since the line is assumed to be electrically short.

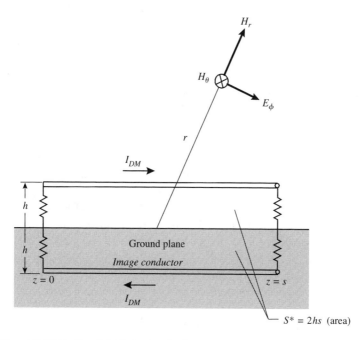

Figure 7.18 Radiated fields from a single conductor over a perfectly conducting ground plane (image formulation).

The conductor and its image in the ground plane are spaced a distance $2h$ apart and form a rectangular loop antenna. This problem is identical to the one shown in Fig. 7.16 for the radiated fields from differential-mode currents on a two-conductor line if the conductor spacing b in Fig. 7.16 is replaced by $2h$.

Then the radiated fields from a conductor over a ground plane are given by (7.97) to (7.103) by replacing

$$S = bs \qquad \text{m}^2 \tag{7.96}$$

with

$$\boxed{S^* = 2hs \qquad \text{m}^2.} \tag{7.113}$$

REFERENCES

[1] C. R. Paul, *Analysis of Multiconductor Transmission Lines*, A Wiley-Interscience Publication, John Wiley & Sons, New York, 1994.

[2] B. C. Wadell, *Transmission Line Design Handbook*, Artech House, Inc., Norwood, Massachusetts, 1991.

[3] W. C. Johnson, *Transmission Lines and Networks*, McGraw-Hill Book Company, New York, 1950.

[4] R. E. Collin, *Field Theory of Guided Waves*, IEEE/OUP Series on Electromagnetic Wave Theory, IEEE Press Selected Reprint Volume, IEEE Press, Piscataway, NJ.

[5] T. Itoh, *Transmission Lines*, Chapter 29 in *Reference Data for Engineers: Radio, Electronics, Computer, and Communications*, Howard W. Sams & Company, Division of Macmillan, Inc., Indianapolis, Indiana, seventh edition, fifth printing, 1989.

[6] R. A. Chipman, *Theory and Problems of Transmission Lines*, Schaum's Outline Series, McGraw-Hill Book Company, New York, 1968.

[7] C. D. Taylor, R. S. Satterwhite, and C. W. Harrison, Jr., "The Response of a Terminated Two-Wire Transmission Line Excited by a Nonuniform Electromagnetic Field," *IEEE Transactions on Antennas and Propagation* (communication), vol. AP-13, no. 6, November 1965, pp. 987–989.

[8] A. A. Smith, Jr., "A More Convenient Form of the Equations for the Response of a Transmission Line Excited by Nonuniform Fields," *IEEE Transactions on Electromagnetic Compatibility*, vol. EMC-15, no. 3, August 1973, pp. 151–152.

[9] A. A. Smith, Jr., *Coupling of External Electromagnetic Fields to Transmission Lines*, first ed., John Wiley & Sons, New York, 1977, second ed., Interference Control Technologies, Gainesville, VA, 1987.

[10] C. R. Paul, "Frequency Response of Multiconductor Transmission Lines Illuminated by an Incident Electromagnetic Field," *IEEE Transactions on Electromagnetic Compatibility*, vol. EMC-18, no. 4, November 1976, pp. 183–190.

[11] A. K. Agrawal, H. J. Price, and S. H. Gurbaxani, "Transient Response of Multiconductor Transmission Lines Excited by a Nonuniform Electromagnetic Field," *IEEE Transactions on Electromagnetic Compatibility*, vol. EMC-22, no. 2, May 1980, pp. 119–129.

[12] F. Rachidi, "Formulation of the Field-to-Transmission Line Coupling Equations in Terms of Magnetic Excitation Field," *IEEE Transactions on Electromagnetic Compatibility*, vol. EMC-35, no. 3, August 1993, pp. 404–407.

[13] F. M. Tesche, M. V. Ianoz, and T. Karlsson, *EMC Analysis Methods and Computational Models*, A Wiley-Interscience Publication, John Wiley & Sons, New York, 1997.

[14] C. R. Paul and D. R. Bush, "Radiated Emissions From Common-Mode Currents," *IEEE Symposium on Electromagnetic Compatibility*, Atlanta, GA, August 1987.

[15] C. R. Paul, "A Comparison of the Contributions of Common-Mode and Differential-Mode Currents in Radiated Emissions," *IEEE Transactions on Electromagnetic Compatibility*, vol. EMC-31, no. 2, May 1989, pp. 189–193.

APPENDIX A

PHYSICAL CONSTANTS

c	wave velocity in free space	$= 1/\sqrt{\mu_0 \varepsilon_0} = 3 \times 10^8$ m/s
μ_0	permeability of free space	$4\pi \times 10^{-7}$ H/m
ε_0	permittivity of free space	8.854×10^{-12} F/m
Z_0	intrinsic impedance of free space	$= \sqrt{\frac{\mu_0}{\varepsilon_0}} = 377$ ohms
e	electron charge	-1.602×10^{-19} C
k	Boltzmann's constant	1.38×10^{-23} J/K (joules/kelvin)
h	Plank's constant	6.626×10^{-34} J $\cdot$ s (joule-sec)
σ_c	conductivity of copper	5.8×10^7 mhos/m

ELECTRICAL UNITS

Q	electric charge	$6.242 \times 10^{18} e = 1\text{A} \cdot \text{s}$	C (coulomb)
I	electric current		A (ampere)
V	voltage		V (volt)
W	energy		J (joule)
P	power		W (watt)
σ	conductivity		S/m (siemens per meter)
ρ	resistivity		Ω m (ohm meter)
ψ	electric flux		C (coulomb)
D	electric flux density	$= \varepsilon E$	C/m^2 (coulomb per sq meter)
E	electric field strength		V/m (volt per meter)
ε	permittivity	$= \varepsilon_o \varepsilon_r$	F/m (farad per meter)
ε_r	relative permittivity		(numeric)
ϕ	magnetic flux		Wb (weber)
B	magnetic flux density	$= \mu H$	T (tesla)
H	magnetic field strength		A/m (ampere per meter)
μ	permeability	$= \mu_o \mu_r$	H/m (henry per meter)
μ_r	relative permeability		(numeric)
ω	radian frequency	$= 2\pi f$	rad/s (radians per second)
f	frequency		Hz (hertz)
f_M	frequency in megahertz		MHz (megahertz)
α	attenuation constant		Np/m (neper per meter)
β	phase constant		rad/m (radians per meter)
γ	propagation constant	$= \alpha + \text{j}\beta$	m^{-1} (per meter)
υ	wave velocity in any medium	$= 1/\sqrt{\mu\varepsilon}$	m/s (meters per sec)
λ	wavelength		m (meters)

APPENDIX C

WAVE RELATIONS

FREE SPACE

$$\beta_0 = \frac{\omega}{c} = \omega\sqrt{\mu_0\varepsilon_0} = \frac{2\pi}{\lambda_0} = \frac{2\pi f}{c} = \frac{\pi f_M}{150}$$

$$\varepsilon_0 c = \sqrt{\frac{\varepsilon_0}{\mu_0}} = \frac{1}{120\pi}$$

$$\mu_0 c = \sqrt{\frac{\mu_0}{\varepsilon_0}} = 120\pi$$

$$\varepsilon_0 \omega = \frac{\beta_0}{120\pi}$$

$$\mu_0 \omega = 120\pi \beta_0$$

$$\frac{\beta_0}{120\pi \omega \varepsilon_0} = \frac{1}{120\pi c \varepsilon_0} = 1$$

$$\frac{120\pi \beta_0}{\omega \mu_0} = \frac{120\pi}{\mu_0 c} = 1$$

ANY MEDIUM

$$Z = \sqrt{\frac{\mu}{\varepsilon}} \qquad \text{intrinsic impedance}$$

$$\beta = \frac{\omega}{v} = \omega\sqrt{\mu\varepsilon} = \frac{2\pi}{\lambda} = \frac{2\pi f}{v}$$

$$\frac{\beta}{Z\omega\varepsilon} = \frac{1}{Z\varepsilon v} = 1$$

$$\frac{Z\beta}{\omega\mu} = \frac{Z}{\mu v} = 1$$

PLANE WAVE IN FREE SPACE

$$E = 120\pi H = \sqrt{\frac{\mu_0}{\varepsilon_0}} H = \frac{B}{\sqrt{\mu_0\varepsilon_0}} = cB = \frac{\omega}{\beta_0} B$$

APPENDIX **D**

MATH IDENTITIES

TRIGONOMETRIC

$$\varepsilon^{jx} = \cos x + j \sin x \qquad \varepsilon^{-jx} = \cos x - j \sin x$$

$$\sin x = \frac{\varepsilon^{jx} - \varepsilon^{-jx}}{2j} \qquad \cos x = \frac{\varepsilon^{jx} + \varepsilon^{-jx}}{2}$$

$$\sin^2 x + \cos^2 x = 1 \qquad \sin^2 x \cos^2 x = \frac{\sin^2 2x}{4}$$

$$\sin^2 x = \frac{1 - \cos 2x}{2} \qquad \cos^2 x = \frac{1 + \cos 2x}{2}$$

$$\sin 2x = 2 \sin x \cos x$$

$$\cos 2x = \cos^2 x - \sin^2 x = 1 - 2\sin^2 x = 2\cos^2 x - 1$$

$$\sin(x \pm y) = \sin x \cos y \pm \cos x \sin y$$

$$\cos(x \pm y) = \cos x \cos y \mp \sin x \sin y$$

HYPERBOLIC

$$\varepsilon^{x} = \cosh x + \sinh x \qquad \varepsilon^{-x} = \cosh x - \sinh x$$

$$\sinh x = \frac{\varepsilon^{x} - \varepsilon^{-x}}{2} \qquad \cosh x = \frac{\varepsilon^{x} + \varepsilon^{-x}}{2}$$

$$\cosh^2 x - \sinh^2 x = 1 \qquad \tanh x = \frac{\sinh x}{\cosh x}$$

$$\sinh 2x = 2 \sinh x \cosh x \qquad \cosh 2x = \cosh^2 x + \sinh^2 x$$

$$\sinh(x \pm y) = \sinh x \cosh y \pm \cosh x \sinh y$$

$$\cosh(x \pm y) = \cosh x \cosh y \pm \sinh x \sinh y$$

TRIGONOMETRIC-HYPERBOLIC RELATIONSHIPS

$$\sinh jx = j \sin x \qquad \sin jx = j \sinh x$$
$$\cosh jx = \cos x \qquad \cos jx = \cosh x$$
$$\tanh jx = j \tan x \qquad \tan jx = j \tanh x$$

$$\sinh(x \pm jy) = \sinh x \cos y \pm j \cosh x \sin y$$
$$\cosh(x \pm jy) = \cosh x \cos y \pm j \sinh x \sin y$$

APPENDIX **E**

VECTOR OPERATORS

The gradient of a scalar ($\nabla\Phi$), the divergence of a vector ($\nabla \cdot \mathbf{F}$), and the curl of a vector ($\nabla \times \mathbf{F}$) are listed below for rectangular, cylindrical, and spherical coordinates.

RECTANGULAR COORDINATES

$$\nabla\Phi = \mathbf{a}_x \frac{\partial\Phi}{\partial x} + \mathbf{a}_y \frac{\partial\Phi}{\partial y} + \mathbf{a}_z \frac{\partial\Phi}{\partial z}$$

$$\nabla \cdot \mathbf{F} = \frac{\partial F_x}{\partial x} + \frac{\partial F_y}{\partial y} + \frac{\partial F_z}{\partial z}$$

$$\nabla \times \mathbf{F} = \mathbf{a}_x \left(\frac{\partial F_z}{\partial y} - \frac{\partial F_y}{\partial z} \right) + \mathbf{a}_y \left(\frac{\partial F_x}{\partial z} - \frac{\partial F_z}{\partial x} \right) + \mathbf{a}_z \left(\frac{\partial F_y}{\partial x} - \frac{\partial F_x}{\partial y} \right)$$

CYLINDRICAL COORDINATES

$$\nabla\Phi = \mathbf{a}_\rho \frac{\partial\Phi}{\partial\rho} + \mathbf{a}_\phi \frac{1}{\rho}\frac{\partial\Phi}{\partial\phi} + \mathbf{a}_z \frac{\partial\Phi}{\partial z}$$

$$\nabla \cdot \mathbf{F} = \frac{1}{\rho}\frac{\partial}{\partial\rho}(\rho F_\rho) + \frac{1}{\rho}\frac{\partial F_\phi}{\partial\phi} + \frac{\partial F_z}{\partial z}$$

$$\nabla \times \mathbf{F} = \mathbf{a}_\rho \left(\frac{1}{\rho}\frac{\partial F_z}{\partial\phi} - \frac{\partial F_\phi}{\partial z} \right) + \mathbf{a}_\phi \left(\frac{\partial F_\rho}{\partial z} - \frac{\partial F_z}{\partial\rho} \right) + \mathbf{a}_z \left(\frac{1}{\rho}\frac{\partial(\rho F_\phi)}{\partial\rho} - \frac{1}{\rho}\frac{\partial F_\rho}{\partial\phi} \right)$$

SPHERICAL COORDINATES

$$\nabla \Phi = \mathbf{a}_r \frac{\partial \Phi}{\partial r} + \mathbf{a}_\theta \frac{1}{r} \frac{\partial \Phi}{\partial \theta} + \mathbf{a}_\phi \frac{1}{r \sin \theta} \frac{\partial \Phi}{\partial \phi}$$

$$\nabla \cdot F = \frac{1}{r^2} \frac{\partial}{\partial r} (r^2 F_r) + \frac{1}{r \sin \theta} \frac{\partial}{\partial \theta} (F_\theta \sin \theta)$$

$$+ \frac{1}{r \sin \theta} \frac{\partial F_\phi}{\partial \phi}$$

$$\nabla \times F = \frac{\mathbf{a}_r}{r \sin \theta} \left(\frac{\partial}{\partial \theta} (F_\phi \sin \theta) - \frac{\partial F_\theta}{\partial \phi} \right) + \frac{\mathbf{a}_\theta}{r} \left(\frac{1}{\sin \theta} \frac{\partial F_r}{\partial \phi} - \frac{\partial}{\partial r} (r F_\phi) \right)$$

$$+ \frac{\mathbf{a}_\phi}{r} \left(\frac{\partial}{\partial r} (r F_\theta) - \frac{\partial F_r}{\partial \theta} \right)$$

FREQUENCY BANDS

ITU* FREQUENCY BANDS

	Designation	Frequency Range
ELF	Extremely low frequency	30–300 Hz
VF	Voice frequency	300–3000 Hz
VLF	Very low frequency	3–30 kHz
LF	Low frequency	30–300 kHz
MF	Medium frequency	300–3000 kHz
HF	High frequency	3–30 MHz
VHF	Very high frequency	30–300 MHz
UHF	Ultra-high frequency	300–3000 MHz
SHF	Super-high frequency	3–30 GHz
EHF	Extremely high frequency	30–300 GHz

*International Telecommunication Union

MICROWAVE/RADAR BANDS

Designation	Frequency Range
L band	1–2 GHz
S band	2–4 GHz
C band	4–8 GHz
X band	8–12 GHz
Ku band	12–18 GHz
K band	18–27 GHz
Ka band	27–40 GHz

INDEX

A

Absorption loss, 25
Absorption, in buildings, 60
AM broadcast signals, 60
Ambient noise surveys, 77
Ambient noise, 116
Ampere's law, 9, 10, 12
  differential form, 17
Ampere, Andre, 9
Amplitude spectrum, 143, 145
Anechoic chamber, 89, 98
Antenna calibration, 94–108
Antenna efficiency, total, 75, 76, 79, 83
Antenna factor
  definition, 71
  electric, 71, 97
  horn antenna, 93
  log-periodic antenna, 91
  magnetic, 71, 97
  and power gain, 83, 84
  and realized gain, 82, 83
  resonant dipole, 89, 90
Antenna impedance, 69, 77, 79
  half-wave dipole, 90
  mismatch factor, 79
  resonant dipole, 89
Antenna noise temperature, 76
Antenna power
  incident, input, and radiated, 80
Antenna power pattern, 72
Antenna reflection coefficient, 77, 79, 83

Antenna resistance, 69
Antenna voltage-standing-wave ratio
  (VSWR), 77, 84
Aperture coupling, 28
Atmospheric noise, 120, 125, 127–130
Attenuation constant
  in metal, 25
  transmission line, 161, 163, 164, 166,
  167, 175
Attenuation, ground wave, 52–57
Available noise power, 112–113
Available noise voltage, 112, 115,
  118–120

B

Backward-traveling wave, 163
Balun transformer, 71, 90, 116
Basic transmission loss, 46
Beamwidth, 78, 92, 93
Biological effects, 77
Biot-Savart law, 7, 12
Boltzmann's constant, 76, 112
Boresight, 91
Boundary conditions
  between two regions, 19–21
  at surface of perfect conductor, 20, 21
Brewster angle, 51
Broadband dipole, 90
Broadband receiver sensitivity, 121, 123,
  124

Broadband receiver sensitivity (*Cont'd*)
  versus noise figure, 123, 124
Broadband response, of network, 153, 154
Broadband signals, 121, 122, 155
Broadside incidence
  two-conductor line, 180, 181
  conductor over a ground plane, 184, 185
Building attenuation, 60–63
Burrows-Gray model, 49, 50, 53, 54

C

Cellular radio, 60
cgs units, 12
Characteristic impedance, 70, 161–175, 177, 179–185, 188
Charge density, 15, 16
  surface, 19
Charge distribution
  line charge, 3, 5, 6
  point charge, 1–3, 6
  surface charge, 3, 5, 6
  volume charge, 3
Charge, electric
  moving, 7
  stationary, 1, 2
Circularly polarized fields, 91
Coaxial line, 158, 167, 188
Common mode currents, 157, 159, 160, 187, 183, 188
Common-mode radiation, 190–193
Component loss, 117
Conducted noise, power-line, 132–135
Conduction current density, 15–17
Conduction current, 9
Conductivity
  of conductors, 167
  of earth (table), 51
  of medium, 15, 16, 19
  of metals, 23
  relative, 23
Conical log periodic antenna, 91
Conservation of energy, 4
Conservative field, 10
Convection current density, 16, 17
Convection current, 9

Coplanar line, 158, 167
Cordless phones, 60
Co-site region, 46, 47, 49
Cosmic background temperature, 125
Cosmic radio noise, 125
Coulomb's law, 1, 2, 6
Coulomb, Charles, 1
Current
  conduction, 9
  convection, 9
  direct, 7
  displacement, 9, 17
  surface, 19–21
Current density
  conduction, 15–17
  convection, 16, 17
  displacement, 17
Current distribution
  on transmission line, 170, 171

D

Declination, 135, 136
Depth of penetration, 25
Detector, 121
  diode, 94
  thermocouple, 94
Deterministic signals, 142
Dielectric constant, 51
  effective, 159
Differential mode currents, 157–160, 178, 183, 188
Differential-mode radiation, 189, 190, 192, 194
Diffraction (spherical earth) region, 47
Diffraction loss
  knife edge obstacle, 63, 64
  round obstacle, 63–65
Diffraction theory, 47
Diffusion, 24, 25
Digital signals, 148
Diode detectors, 94
Dip angle, 135
Dipole antenna
  broadband, 90
  half-wave, 89, 90
  receiving, 33, 89, 90
  resonant, 88
  short—*see* Short dipole

tunable, 89, 90
transmitting, 57
Dipole, electric, 33
Direct current, 7
Direct wave, 48, 50, 57, 58
Directive gain
    definition, 73
    and effective length, 82
    of half-wave dipole, 57
    and power density, 80
    and power gain, 78
    of radiator, 45, 56
    and realized gain, 78
    of short dipole, 57, 60, 74
Directivity, 57–60, 78
    definition, 74
    and effective length, 81, 82
    and maximum effective aperture,
        79, 81
    and power gain, 79, 100
    and realized gain, 79
    of short dipole, 74
Discone antenna, 93
Displacement current density, 17
Distributed parameters
    (transmission line), 161–167
Diurnal variation, atmospheric noise, 129
Dry sand, 51

E

Earth's electric field, 137, 138
Earth's magnetic field, 135–137
Earth's magnetic poles, 135–137
Earth-space propagation, 43, 44
Edge diffraction, 63–66
Edge reflections, ground screen, 104
Edge-fire incidence
    conductor over a ground plane,
        185–187
    two-conductor line, 180, 182
Effective antenna noise factor, 112–115,
    117
    of atmospheric noise, 127, 128
    of galactic noise, 131
    of man-made radio noise, 130, 131
Effective antenna noise temperature,
    112, 113

Effective aperture
    definition, 76
    and directivity, 79
    and maximum effective aperture, 79
    and power gain, 79, 80
    and realized gain, 79
Effective height, receiving antenna, 71
Effective length, receiving antenna,
    69–71, 78, 89, 90
    and directive gain, 82
    and directivity, 81, 82
    of small loop, 88
    of standard antenna, 96
E-field probe, 39
Electric dipole, 33
Electric displacement, 3
Electric field strength
    and power density, 80
    static, 2, 3, 4–6
    time-varying, 4, 15–18
Electric field, earth's, 137, 138
Electric flux density, 2, 3, 6
Electric flux, 3, 4, 6
Electromagnetic interference (EMI), 77
Elliptically polarized fields, 91
End-fire incidence
    conductor over a ground plane,
        183, 184
    two-conductor line, 180
Energy density spectrum, 143
Energy signals, 141, 142
Envelope detection, 144
External noise factor, antenna, 76, 78
External noise, 116, 118
Extraterrestrial noise, 125, 126, 132

F

Fall-off
    line charge, 5
    line current, 8
    point charge, 2
Far field, 28, 30, 31, 32–38, 190, 191
    power density, 38
Faraday's law, 4, 13–15
    differential form, 17
Faraday, Michael, 13
Field strength meter, 95, 100, 102, 129,
    132, 144, 146, 153

Flare, solar, 126
Flux
  electric, 3, 4, 6
  magnetic, 8, 9, 12
Flux density
  electric, 2, 3, 6
  magnetic, 8, 9, 12
FM broadcast signals, 60
Force
  and Coulomb's law, 1
  Lorentz, 10, 11
Forward-traveling wave, 163
Fourier series, 142
Fourier transform, 142, 143
Fraunhofer region, 35
Free space propagation, 43–46
Free-space wave-number, 45
Frequency spectrum, 151–153
Fresh water, 51
Fresnel ellipse, 106
Fresnel region, 34

G

Galactic noise, 125–128, 131, 132
Galactic plane, 125, 132
Gauss (cgs unit), 12
Gauss's law 2, 3, 6, 10
  differential form, 17
Gauss, Karl, 2
Gaussian noise, 133
Geomagnetic field, 135–137
Geometric mean, 106
Good earth, 51–57
Ground plane, 57, 99, 101, 102, 104
Ground reflected wave, 58
Ground screen edge reflections, 104
Ground wave field strength, 49, 50,
    104, 106
Ground wave propagation regions, 47
Ground wave propagation, 43, 44,
    46–57, 127
Ground wave, components of, 48
Ground-reflected wave, 48, 50

H

Half-wave dipole, 57, 58, 89, 90
Hazardous radiation, 38–40, 94

Height gain factor, 49
Height gain, ground wave, 52–57
H-field probe, 39
Horn antenna, 92, 93
  standard-gain, 98

I

Ideal receiver
  noise figure, 132
Ideal site, 103, 106
Impedance mismatch loss, 73–75
Impedance, antenna, 69, 77, 79
  half-wave dipole, 90
  mismatch factor, 79
  resonant dipole, 89
Impulse
  definition, 146
  spectral intensity, 146, 147
Impulse bandwidth, 144
  definition, 146
Impulse generator, 146
Impulse strength, 146
Impulsive signals, 121
Incident wave, 21, 22, 24, 26
Inclination, 135, 136
Intrinsic impedance
  of free space, 22, 45
  of medium, 23
  of metals, 23
Ionospheric propagation, 43, 44, 127
ISM (Industrial, Scientific and Medical),
    94
Isotropic radiator, 43, 50, 73

J

Jupiter, noise spectrum, 125, 126

L

Lenz's law, 14
Lightning, 60, 127
Line charge, 3, 5, 6
Line spectrum, 142
Linearly polarized fields, 91
Line-of-sight region, *see* Plane
    earth region
Log normal distribution, 60
Log periodic antenna, 90, 91, 107, 108

Loop antenna
  induced voltage and Faraday's law, 14
  radiated fields, 28, 31–33
  receiving, 33, 87–89
  wave impedance, 37
Loop, magnetic, 33
Lorentz force, 10–12
Lorentz, Hendrick, 10
Loss factor
  antenna, 116
  transmission line, 116, 119
Loss resistance, antenna, 69
Loss tangent, 167
Lunar noise, 125, 126

M

Magnetic equator, 135, 136
Magnetic field strength
  from line current, 8
  static, 7, 8
  time-varying, 14, 17, 18
Magnetic field, earth's, 135–137
Magnetic flux density, 8, 9, 12
Magnetic flux, 8, 9, 12
Magnetic loop, 33
Magnetic poles, earth's, 135–137
Man-made noise, 120, 125, 130–132
Maximum effective aperture
  definition, 76
  and directivity, 79, 81
  and effective aperture, 79
  and power gain, 79
  and realized gain, 79
  of short monopole, 114
Maximum permissible exposure (MPE),
  38–40
Maxwell (cgs unit), 12
Maxwell's equations
  region with sources, 15–17
  sinusoidal fields, 18
  source-free region, 17
Maxwell, James Clerk, 15
Micropulsations, 137
Microstrip line, 158, 167
Microwave fields, 60
Microwave oven (field regions), 35
Milky way galaxy, 125, 132
Mismatch factor, impedance, 79

Mismatch loss, impedance, 73–75
Monopole antenna, 28, 53, 92, 129
Moon, noise spectrum, 125, 126
Multiconductor transmission line, 157,
  158, 167, 176
Multistory office buildings,
  attenuation of fields in, 60–62

N

Narrowband receiver sensitivity, 121,
  123, 124
  versus noise figure, 123, 124
Narrowband response, of network, 153,
  154
Narrowband signals, 121
National Institute of Standards
  and Technology, 57, 95–99, 101
Near field
  radiating, 33–36
  reactive, 28, 30–36, 38, 39, 50, 53,
    190, 192
  wave impedance, 38, 39
Neper, 163, 164
Network analyzer, 144
NIST, 57, 95–99, 101
Noise bandwidth, receiver, 76
Noise factor
  antenna circuit, 116
  cascaded networks, 117
  external, of antenna, 76, 78
  receiver, 117, 118
  system, 116–118
  transmission line, 116
Noise field power spectral density,
  112–114, 118–120
  of atmospheric noise, 127
  of extraterrestrial noise, 125–126
  of man-made radio noise, 130
Noise field spectral intensity, 112–114,
  119, 120
  of atmospheric noise, 127, 129, 131
  of extraterrestrial noise, 125, 126
  of galactic noise, 131
  of man-made radio noise, 130–132
Noise field strength, 112–115
Noise figure, 93, 122, 123–125, 130
  ideal receiver, 132
  and receiver sensitivity, 123, 124

Noise parameters (table), 112
Noise power
  see Received noise power
Noise power spectral density, 112,
  113, 142
  of power-line conducted noise,
    133, 134
Noise temperature, antenna, 76
Noise voltage
  see Received noise voltage
Noise voltage spectral intensity, 112,
  115, 116
  of power-line conducted noise,
    133, 134
Nonuniform transmission line, 158

O

Observable noise power, 118
Obstruction-free area, 106
Oersted (cgs unit), 12
Omnidirectional antenna, 93
Open-circuit voltage, receiving antenna,
  69
Open-field site, 99, 101

P

Parabolic antenna, 93
Parallel polarized wave, 58
Parallel-strip transmission line, 98
Path loss measurements, 77
Path loss, radio, 63
PCS, 60
Perfectly conducting plane
  definition, 57
  propagation over, 57–60
Periodic signals, 142, 151, 153
Permeability
  of conductors, 167
  of free-space, 8, 12
  of medium, 8, 15, 16, 19, 165
  of metals, 23
  relative, 8, 23, 167
Permittivity
  of earth (table), 51
  of free space, 2, 5, 51
  of medium, 2, 15, 19, 165, 167
  relative, 2, 159, 167

Personal computer (field regions), 35, 36
Per-unit-length parameters, 162, 166
  see Distributed parameters
Phase constant
  in metal, 25
  transmission line, 161, 163, 164, 166,
    175, 179
Phase velocity, 161, 164–167, 175, 179
Phasor, 38
Planar log periodic antenna, 91
Plane earth (line-of-sight) region, 47, 48,
  56, 58
Plane of incidence, 57, 58
Plane wave, 34
  power density, 38, 39
  wave impedance, 21–23, 37
Planetary noise, 126
Point charge, 1–3, 6, 10
Potential difference, 4–6
Power density
  average, 38–40
  and directive gain, 80
  free space propagation, 45, 46
  instantaneous, 38
  and power gain, 80
  radiated, 72
  and realized gain, 80
Power distribution systems, 5, 133
Power gain
  and antenna factor, 83, 84
  definition, 74
  and directive gain, 78
  and effective aperture, 79, 80
  and maximum effective aperture, 79
  and power density, 80
  and realized gain, 78, 79
Power pattern, antenna, 72
Power signals, 141, 142
Power spectral density
  and Noise field power spectral density
  see Noise power spectral density
Power, antenna
  incident, input, and radiated, 80
Power-line conducted noise, 132–135
Poynting vector
  average, 38, 72
  instantaneous, 38
Probe,
  E-field, 39

H-field, 39
    radiation monitor, 94
    surface current, 21
Propagation constant
    in metal, 25
    transmission line, 161–163, 166
Propagation modes (table), 44
Pseudo Brewster angle, 51
Pyramidal horn, 92

Q

Quasi-TEM solution, 159
Quiet sun, 126

R

Radar fields, 60
Radiating near-field, 33–36
Radiation efficiency, 75, 78, 79
Radiation hazard meters, 39
Radiation intensity, 72–74
Radiation monitors, 94
Radiation resistance, 69, 75
Radiation, hazardous, 38–40, 94
Radio paging, 60
Radio path loss, 63
Random signals, 142
Ray theory, 50, 53
Rayleigh roughness criterion, 65–67, 106
Reactive near-field, 28, 30–36, 38,
        39, 50, 53
    and power density, 38, 39
Realized gain
    and antenna factor, 82, 83
    definition, 74, 75
    and directive gain, 78
    and effective aperture, 79
    and maximum effective aperture, 79
    and power density, 80
    and power gain, 78, 79
Received noise power
    power-line conducted noise, 133
Received noise voltage, 119, 120, 132
    atmospheric noise, 129
    extraterrestrial noise, 126
    man-made noise, 132
    power-line conducted noise, 133
Received noise, 116, 118–120

Received voltage, antenna, 70
    loop antenna, 15
Receiver noise bandwidth, 76
Receiver noise figure, 93, 122–125, 130
    ideal receiver, 132
    and receiver sensitivity, 123, 124
Receiver noise, 116, 118, 131
Receiver sensitivity, 119–125
    broadband, 121, 123, 124
    definition, 120
    narrowband, 121, 123, 124
Receiver, ideal
    noise figure, 132
Receiving antenna
    effective height, 71
    effective length, 69–71, 89, 90
    equivalent circuit, 70, 85
    open-circuit voltage, 69
    types of, 87–94
Receiving system, 116–125
Reciprocity, 84–87, 188
Rectangular pulse, 147, 149
Reference temperature, 113, 116
Reflected wave, 21, 22
Reflection coefficient
    antenna, 77, 79, 83, 84
    for good earth, 51, 52
    for plane earth, 50, 51
    transmission line, 161, 168–170, 176
    for wave incident on conducting half
space, 21, 22
Reflection loss, 24
Reflection, specular, 65, 66
Reflections, at building walls, 60
Reflector antennas, 93
Refracted wave, 21, 26
Resistance, antenna, 69
Resonant dipole, 89
Right-hand rule, 8
Robert's dipole, 90
Rod antenna, 92
Rough surface, 66

S

Scattered wave, 65, 66
Scattering, in buildings, 60
Schumann resonances, 137
Sea water, 51

Secondary standard-antenna method, 95,
    99–101
Semi-anechoic chamber, 57
Sensitivity
    broadband, 121, 123, 124
    definition, 120
    narrowband, 121, 123, 124
    receiver, 119–125
Shielded pair, 158, 167
Shielded room, 89
Shielding effectiveness, 26–28
Shielding factor, 26, 28
Short dipole
    directive gain, 57, 59, 74
    directivity, 74
    radiated fields, 28–32, 53, 59
    wave impedance, 37
Short monopole, 28, 92
    maximum effective aperture, 114
Short-wave receiver, 129
SI units, 12
Signal generator, 95, 102
Signal-to-noise ratio, 121
Single family residences,
    attenuation of fields in, 60, 62
Single-story concrete block buildings,
    attenuation of fields in, 60, 62
Site attenuation, 57, 101, 103, 107
Skin depth, 20, 24, 25
Sky wave propagation, 43, 44
Small loop antenna
    radiated fields, 28, 31–33
    receiving, 33, 87–89
    wave impedance, 37
Smooth surface, 67
Solar flare, 126, 10
Solar noise, 125, 126
Space vector, 38
Space wave, 48, 49, 53, 54
    over perfectly conducting plane, 57
Spectral intensity, 143–151
    of impulse, 146
    of rectangular pulse, 147–149, 154
    of trapezoidal pulse, 148–151
Spectral lines, 151–154
Spectrum analyzer, 95, 100, 102, 107,
    129, 132, 144, 153

Specular reflection, 66
Spherical earth (diffraction) region, 47
Spherical wave, 43, 45
Standard-antenna method, 95–97
Standard-field method, 95, 97–99
Standard-gain horn, 98
Standard-site method, 95, 97, 101–108
Step discontinuity, 66
Strip line, 158, 167
Sun, noise spectrum, 125, 126
Sunspot, 126
Surface charge density, 19
Surface charge, 3, 5, 6
Surface current probe, 21
Surface current, 19–21
Surface roughness, 66, 106
Surface wave attenuation factor, 49
Surface wave, 48, 49, 53, 54, 107
    over perfectly conducting plane, 57
System noise factor, 116–118

T

TE wave, 36, 159
Telegrapher's equations, 162
TEM cell, 98, 99
TEM mode, 157–159
TEM wave, 23, 36, 158
Temperature
    antenna circuit, 116
    cosmic background, 125
    reference, 113, 116
    transmission line, 116
Tesla (SI unit), 12
Thermocouple detectors, 94
Thunderstorms, 127
TM wave, 36, 159
Total antenna efficiency, 75, 76, 79, 83
Tracking generator, 95, 102, 107
Transmission coefficient, 21–24, 26, 27
    transmission line, 161, 168–170
Transmission line
    loss factor, 116, 119
    temperature, 116
Transmitted wave, 21, 22 ,24, 26
Transmitting antenna
    dipole, 57
    equivalent circuit, 70, 85

Transverse electric (TE) wave, 36, 159
Transverse electromagnetic (TEM) cell,
  98, 99
Transverse electromagnetic (TEM) mode,
  157–159
Transverse electromagnetic (TEM) wave,
  23, 36, 158
Transverse magnetic (TM) wave, 36, 159
Trapezoidal pulse, 148–151
Tropospheric scatter propagation, 43,
  44, 47
Tunable dipole, 89–90
TV broadcast signals, 60
Two-wire line, 158, 167

U
Uniform transmission line, 158

V
Voltage distribution
  on transmission line, 170, 171

Voltage spectral intensity
  see Noise voltage spectral intensity
Voltage standing-wave ratio (VSWR)
  antenna, 77, 84
  transmission line, 174–176
Volume charge, 3

W
Wave impedance
  and characteristic impedance, 165
  of electromagnetic wave, 36, 165
  inside buildings, 62
  near field, 38, 39
  of plane wave, 21–23, 37
  of short dipole, 37
  of small loop, 37
Wave number, free space, 45
Waveguide modes, 159
Weber (SI unit), 12
Wire over ground, 158
Wireless communication signals, 60

ABOUT THE AUTHOR

Albert A. Smith, Jr. received the B.S.E.E. degree from the Milwaukee School of Engineering in 1961 and the M.S.E.E. degree from New York University in 1964. From 1961 to 1964 he was employed by the Adler-Westrex Division of Litton Industries where he was engaged in the design of HF and Troposcatter communication systems.

In 1964 he joined the IBM Electromagnetic Compatibility Laboratory in Kingston, New York. During his career at IBM, Mr. Smith performed theoretical and experimental investigations in areas such as wave propagation, antennas, radio noise, lightning effects, electromagnetic data security, TEMPEST, and the biological effects of non-ionizing radiation. As a senior engineer at the IBM EMC Laboratory in Poughkeepsie, New York, he was responsible for the electromagnetic compatibility of IBM's large computer systems. He designed the IBM Poughkeepsie Open Area Test Facility for measuring electromagnetic radiation from large computer systems, the largest facility of its kind in the world with a 50-ft diameter turntable, a 70 × 70-ft plastic building, and a 100 × 180-ft ground screen.

Mr. Smith has made fundamental contributions to the theory of the coupling of electromagnetic fields to transmission lines, the characterization of electromagnetic measurement sites, and antenna calibration. He has contributed to the development of national and international EMC standards.

A Fellow of the IEEE, Mr. Smith is associate editor of the *IEEE Transactions on Electromagnetic Compatibility* and past chairman of the IEEE Technical Committee on Electromagnetic Environments. Also, Mr. Smith has served on American National Standards Committee C63.1, Techniques and Developments, and Computer Business Equipment Manufacturers Association subcommittee on Electromagnetic Compatibility, ESC-5. He is the author of the book *Coupling of External Electromagnetic Fields to Transmission Lines* and coauthor, with Edward N. Skomal, of *Measuring the Radio Frequency Environment*. A member of Eta Kappa Nu, Tau Omega Mu, and Kappa Eta Kappa, Mr. Smith has published more than 20 technical papers.